T0254023

# GIS-Based Simulation and Analysis of Intra-Urban Commuting

# GIS-Based Simulation and Analysis of Intra-Urban Commuting

Yujie Hu
Fahui Wang

**CRC Press**
Taylor & Francis Group
Boca Raton London New York

CRC Press is an imprint of the
Taylor & Francis Group, an **informa** business

CRC Press
Taylor & Francis Group
6000 Broken Sound Parkway NW, Suite 300
Boca Raton, FL 33487-2742

First issued in paperback 2020

© 2019 by Taylor & Francis Group, LLC
CRC Press is an imprint of Taylor & Francis Group, an Informa business

No claim to original U.S. Government works

ISBN-13: 978-0-367-02303-4 (hbk)
ISBN-13: 978-0-367-60659-6 (pbk)

**Visit the Taylor & Francis Website at**
**http://www.taylorandfrancis.com**

**and the CRC Press Website at**
**http://www.crcpress.com**

*Yujie Hu would like to dedicate the book in loving memory of Peiyu Cong (1935–2003) and Jianshui Yu (1932–2018) and to Jing and our son Ben*

# Contents

# Preface

PEOPLE TRAVEL FOR VARIOUS purposes, and for most, their daily journey-to-work trips are the most consistent. It is widely believed that residents consider the duration of commuting as the primary factor in residential choice. Therefore, this daily link between residences and workplaces sets up the complex interaction between the two most important land uses (residential and employment) in a city and dictates the configuration of urban structure. Commuting also reflects one's job accessibility. The most congested periods in a day occur during the morning and afternoon rush hours. In addition to prolonged time and stress for individual commuters in traffic, commuting comes with additional societal costs including elevated crash risks, worsening air quality, louder traffic noise, and so on. These issues are important to city planners, policy researchers, and decision makers. If you are interested in mapping, understanding, predicting, and planning journey-to-work trips, this is the book for you.

As geographers, we are always fascinated by how people and places interact and are organized. We travel, and by traveling, we connect to each other at different places. Geographic models help predict where and when we travel, and is termed human mobility. Individual travels collectively form patterns. The advancement of geographic information systems (GIS) has enabled our discovery of these patterns more precisely and more efficiently. Commuting, perhaps the most prevalent trip purpose in a city, has long been studied in various disciplines such as geography, urban planning, economics, and transportation engineering. But few books focus on how to use GIS to map, discover, analyze, and understand commuting patterns. This book reports our most recent efforts on this topic. We employ a Monte Carlo simulation approach to model individual commuting trips in order to gain better estimates and understanding of trip lengths and assess the potential of reducing them.

The book is a spin-off from Yujie Hu's dissertation, which was completed under Fahui Wang's advisorship and successfully defended in the summer of 2016. Some of the results from the dissertation have been reported in several journal articles (Hu and Wang, 2015a, 2016; Hu et al., 2017) and a book chapter (Hu and Wang, 2015b). However, the rich and important technical details about how the analyses were implemented are left out in these publications. The methods developed in the dissertation have also been applied and refined in other studies that demonstrate their value beyond commute research and merit discussion here. A book also gives us the opportunity to provide a complete and synthesized perspective. With this in mind, we extend our previous work by adding more technical details and some more recent applications in the book. All technical steps are illustrated in step-by-step case studies based on the original data. All graphics are redone for improved quality.

The book consists of seven chapters. Chapter 1 is an overview of the subject matter. Chapter 2 reviews existing commuting studies and covers four areas: measuring commuting lengths, understanding commuting patterns by spatial and nonspatial factors, respectively, and estimating wasteful commuting. Chapter 3 describes the study area and data sources for the case studies throughout the book. Chapter 4 explains the basics of Monte Carlo simulation and its common applications in geographic studies. Also included in Chapter 4 are the principles of two Monte Carlo–based processes—one to simulate individual trip ends and the other to simulate individual trips—and corresponding technical details showcased in case studies using ArcGIS. Chapter 5 describes techniques for measuring commuting lengths based on the simulated trips in Chapter 4, presents a temporal analysis of commuting patterns between 1990 and 2010, and explains the observed commuting patterns by three land use indices. Chapter 6 examines the proposition of wasteful commuting from a perspective of commuting efficiency. It covers two conventional models in the literature and our own model, each illustrated in case studies. Chapter 7 summarizes the major findings, methodological contributions, and limitations of the book. It also discusses how big data may further advance future work on commuting.

This book intends to serve students, researchers, and practitioners in geography, urban planning, public policy, transportation engineering, and other related disciplines. Many students think of GIS as only a mapping tool. GIS is far more than that, however. One reason that prevents students from appreciating its full capacity is the lack of easy-to-follow case studies built upon real-world data. Such technical details are usually absent in academic journal articles. It is our hope that students and junior scholars will be able to replicate the analyses described in the book and beef up their related analytical skills, especially in transportation GIS (GIS-T). All data and computer programs are available for download from our websites: http://faculty.cas.usf.edu/yhu/ (Hu) and http://ga.lsu.edu/blog/fahui/ (Wang).

The authors have received countless assistance from many quarters. Our appreciation first goes to Dr. Chester Wilmot, Dr. Lei Wang, and Dr. Wesley Shrum at Louisiana State University who graciously served as Hu's dissertation committee and provided invaluable guidance. Dr. Hu would like to thank the School of Geosciences, University of South Florida (USF), for the teaching release that made the timely completion of this book possible. He appreciates the support and encouragement from his colleagues at USF, especially Joni Firat, Steven Reader, and Ruiliang Pu. We thank Irma Shagla Britton, Senior Editor for Environmental Sciences, GIS & Remote Sensing, with the CRC Press/Taylor & Francis, for reaching out to us on the short-form publishing initiative, CRC Focus books.

# Authors

Yujie Hu, PhD, is Assistant Professor of GIScience in the School of Geosciences at the University of South Florida in Tampa, Florida. He was a Research Fellow (2016–2017) with the Kinder Institute for Urban Research at Rice University in Houston, Texas. His research and teaching interests focus on the development and applications of GIS, spatial analysis, and visualization techniques in urban systems and public policy implications. The subjects of his research include transportation, public health, crime, and human–environment interactions. His research has been covered by *ABC 13 News* of Houston and the Houston Public Media News, and supported by the Florida Sea Grant, Southeastern Association of State Highway and Transportation Officials (SASHTO), and the American Association of Geographers (AAG). He is a board member of the Transportation Geography Specialty Group of the American Association of Geographers.

Fahui Wang, PhD, is James J. Parsons Professor and Chair of the Department of Geography and Anthropology, Louisiana State University (LSU). His research focuses on applications of GIS and computational methods in human geography (including urban, transportation, economic, cultural, and historical geography)

and public policy (including urban planning, public health, and public safety). His work has been supported by the National Science Foundation, U.S. Department of Energy, U.S. Department of Health and Human Services (Agency for Healthcare Research & Quality and the National Cancer Institute), U.S. Department of Housing and Urban Development, U.S. Department of Justice (National Institute of Justice and Office of Juvenile Justice and Delinquency Prevention), and the National Natural Science Foundation of China. He has published over 130 refereed articles and five books (including two edited books). He was a recipient of the LSU Rainmaker Awards for outstanding research, scholarship, and creative activity (2009, 2015) and the LSU Distinguished Faculty Award (2018), and a Policy Winner of the 2015 Outstanding Article of the Year from the Agency for Healthcare Research and Quality, U.S. Department of Health and Human Services.

# List of Datasets and Program Files

| Section | Steps | Dataset | Program File |
|---|---|---|---|
| 4.3 | 1, 2 | `trt, trtpt` | `Features To Text File.tbx` |
| 4.3 | 3 | `trt.txt,`[a] `wrk_job_2010. txt`[a] | `TSME.exe` |
| 4.4 | 4 | `simu_Os.txt,`[a] `simu_ Ds.txt,`[a] `trip_mtx.txt`[a] | `TSME.exe` |
| 5.1 | 3, 4 | `rd_2010, simu_trips.txt`[a] | |
| 5.1 | 5 | `ctppTT, ctppFlow` | |
| 5.1 | 6 | `ctppTT_Flow`[a] | |
| 5.3 | 10 | `BRCenter, trtpt` | |
| 5.3 | 11 | `Mean_Comm_Dist,`[a] `Mean_Comm_Time,`[a] `Dist_from_CBD`[a] | |
| 5.4 | 13 | `Mean_Comm_Dist,`[a] `Mean_ Comm_Time,`[a] `JWR_5mile`[a] | |
| 5.5 | 15 | `trtpt, ctppFlow_Dist`[a] | |
| 5.5 | 17 | `Mean_Comm_Dist,`[a] `Mean_ Comm_Time,`[a] `JobP_Dist`[a]`, JobP_Time`[a] | |
| 6.3 | 1, 2, 3 | `trt, trtpt, trt_MBC`[a] | |
| 6.3 | 6 | `wrk_job_inBR_2010.txt, odtime.txt,`[a] `oddist.txt`[a] | `WasteComm_LP.R` |
| 6.4 | | `trt.txt,`[a] `wrk_job_ inBR_2010.txt` | `TSME.exe, WasteComm_ILP.R` |

[a] Denotes the datasets that are to be produced in case studies; they are provided for convenience.

# CHAPTER 1

# Introduction

COMMUTING TAKES PLACE ON a daily basis by multiple transportation modes such as driving alone, carpooling, public transit, biking, and walking. By linking home (residential areas) to employment (commercial, industrial, and other land uses), commuting is a key factor in affecting land use patterns (American Association of State Highway and Transportation Officials, 2013). As population and jobs become increasingly decentralized, commuters are traveling longer (both length and duration) than ever before (Gordon et al., 2004). For example, the average commute time by the U.S. worker for their one-way trips to work was 24 minutes in 2003, and increased to 25.1 minutes by 2009 (McKenzie and Rapino, 2011). According to the National Household Travel Survey, one-out-of-twelve U.S. workers spent an hour or more on their one-way commute trips in 2001, a significant increase from one-out-of-twenty in 1995. This is especially worse in large cities where one-out-of-ten workers commuted an hour or more one way for an average of about 38 miles (National Household Travel Survey, 2006). In the meantime, an increasing number of commuters are choosing to drive alone, especially during the present job decentralization era where inter-suburb and reverse commuting are more common. For instance, 75.5 percent of U.S. workers drove alone to jobs in 2008, and the ratio increased to 76.1 percent in 2009 (Fields and Jiles, 2009).

Given the increasingly longer length and duration and the dominantly auto-dependent travel mode, commuting is strongly connected with some practical issues on which many public policies concentrate (Sultana and Weber, 2014). For example, it is a major contributor to traffic congestion, air pollution, and greenhouse gas emissions. Even though commuting represents only a 20–25 percent share of all-purpose trips in the United States (Sultana, 2002; Horner, 2004), it results in two of the most congested periods in a day and establishes the major transportation infrastructure and service needs. For instance, the average commute speed declined about 10 percent in midsize metropolitan areas from 1990 to 2009 (Santos et al., 2011). In line with the worsening traffic congestion, commuters are spending more time on their daily commute trips all across the United States. According to the Texas Transportation Institute (2011), the average commuter spent an additional 34 hours annually and wasted 14 gallons of gas sitting in traffic in 2010. In addition, according to the U.S. Environmental Protection Agency (2018), transportation (including commuting) is "the largest share of greenhouse gas emissions" among all economic sectors in 2016. According to the Highway Statistics (U.S. Department of Transportation/Federal Highway Administration, 2012) and Statistical Abstract of the United States (U.S. Census Bureau, 2012), the Vehicle Miles Traveled (VMT) in U.S. urban areas increased by 133 percent from 1980 to 2012 while the population increased by 36 percent during the same period. Therefore, efforts that focus on studying the urban commuting patterns, for example, understanding the temporal change of commuting and its underlying causes, contribute to reaching the larger goals of traffic congestion mitigation and carbon emission control.

Commuting has been widely studied in previous research such as measuring commuting lengths, wasteful commuting, jobs–housing balance and job accessibility, and investigating the influences of the urban form on commuting patterns (Cervero, 1989; Horner, 2004; Sultana, 2002; Sultana and Weber, 2014; Wang, 2001). However, most studies are affected by the aggregation errors (e.g., measuring commuting distances between centroids of zones) and scale effect (e.g., inconsistent analysis results due to different unit scales and unit zone definitions). More accurate measures of commuting lengths and patterns remain very much needed.

The very focus of this book is to design and apply GIS-based methods for detecting and analyzing intra-urban commuting patterns. The remainder of the book is structured as below.

Chapter 2 reviews existing commuting studies in four areas: measuring commuting lengths, explaining commuting by spatial factors, explaining commuting by nonspatial factors, and measuring wasteful commuting. The literature review heightens the need for more accurate measures of commute distance and time, which forms the core objective of this book.

Chapter 3 describes the study area and data sources used in the book. East Baton Rouge Parish in Louisiana (a parish is a county-equivalent unit in Louisiana) is selected as the study area in all case studies. Major data sources used in the book include census unit boundaries such as census tracts and road networks in a GIS format and commuting and socioeconomic data such as the number of resident workers and the number of employment in a tract and the number of commuters between two tracts in a text format from the Census for Transportation Planning Packages (CTPPs).

Chapter 4 proposes simulation-based approaches and demonstrates their applications in measuring commuting patterns. Specifically, it first gives a brief introduction to the methods and then illustrates the principles and processes of two applications—one to simulate locations of trip ends (resident workers and jobs) and the other to simulate individual commuting trips between them. The simulations enable more accurate measures of commute distance and time, and thus mitigate aggregation errors in area unit and scale effects. A user-friendly program, termed

"Traffic Simulation Modules for Education (TSME)," has been developed to automate both tasks.

Chapter 5 demonstrates the analyses of measuring commuting lengths (in both distance and time) using the simulation methods explained in Chapter 4. It then examines the temporal trends of commuting patterns from 1990 and 2010 in the study area. Finally, it describes several statistical analyses that investigate the relationships between commuting and each of three selected land use indices—distance from the central business district (CBD), jobs–housing balance ratio, and proximity to jobs. The results lend support to the promise of planning policies that aim at trip reduction by improving the jobs–housing balance and job proximity.

Chapter 6 deals with the commuting efficiency (wasteful commuting) issue, another classic topic in commuting research. This study improves over conventional models by proposing an integer linear programming approach to the calibration of minimum commuting. As the measure of wasteful commuting is now at the most disaggregate level (individuals), it is free from any zonal or scale effect present in traditional studies based on an area unit.

Chapter 7 summarizes the major findings, methodological contributions, and limitations of the book. It also discusses future possibilities of employing big data in commuting research.

Readers may follow the detailed step-by-step instructions in Chapters 4 through 6 to replicate the case studies with provided data and computer programs (downloadable via the links provided in the Preface) and apply the methods in designing their own studies.

# CHAPTER 2

# Literature Review

C OMMUTING IS A DYNAMIC process, and the patterns usually vary in different times, places, and socioeconomic groups. For example, commuting around 8:00 a.m. could be more remarkable than around 10:00 a.m.; commuting originating from places far away from the downtown area could be dissimilar to places near the central business district (CBD); and the commuting behavior and patterns may not be the same for the White vs. African American (or male vs. female) population. Given its complex nature and its important role in our daily lives, this subject has received considerable attention in the literature across a wide range of disciplines including geography, urban economics, urban planning, and transportation engineering. Existing studies, particularly in geography, have focused on measuring, modeling, and explaining commuting patterns in order to gain a better understanding of this particular purpose of trips (and the more general human mobility) and its social and environmental impacts. This chapter summarizes existing commuting studies into the following aspects: Section 2.1 discusses previous work in measuring commuting lengths and patterns; Section 2.2 reviews studies that examine the spatial components of commuting; Section 2.3, on the other hand, focuses on the socioeconomic components of commuting; and Section 2.4 examines another line of commuting research—commuting efficiency such as the so-called wasteful commuting.

## 2.1 MEASURING COMMUTING LENGTHS

The topic of measuring commuting lengths deserves some discussion as it is the core of this book. In the commuting literature, time is often utilized to analyze the spatial separation between home and job sites, since it is directly available from survey data. However, mileage could provide a more consistent measure of commuting lengths (Sultana and Weber, 2007) and is not studied as much as time. To measure commuting lengths in terms of distances, existing studies commonly estimated the Euclidean distances between centroids of zones (Gera, 1979; Hamilton, 1982; Levinson and Kumar, 1994; Levinson, 1998; Horner and Murray, 2002; Clark et al., 2003; Wang, 2003; Kim, 2008). However, some have argued that estimates from this centroid-to-centroid straight-line distance measure are not close to the actual figures because this method lacks the consideration of the real transportation network. Hence, they recommended using the travel distance through the road network. Specifically, this distance metric estimates the commuting distance as the shortest path distance between two zone centroids using GIS network modeling techniques (White, 1988; Levinson and Kumar, 1994; Cervero and Wu, 1998; Wang, 2000, 2001; Horner, 2002; Yang, 2008). Clearly, this method assumes that commuters choose the shortest routes to travel to work. While not always true, it is a much closer representation of actual commute distances, especially compared to the Euclidean distances. Though more complex, this distance measure, however, could underestimate the actual distances, since it is still a measure between zone centroids where all people are assumed to start and end a journey. Such an approach could bias the estimates, particularly in large zones, by omitting intrazonal distances. A complete measure of the commuting trip includes the intrazonal ones at both ends and the interzonal one between the two ends' centroids.

As Hewko et al. (2002) noted, aggregation errors resulting from using a centroid to represent a neighborhood in measuring the distance would be significant for analyses based on aggregated units such as census tracts. For this reason, Horner and Schleith (2012) used a small analysis unit, census blocks, in order to reduce such aggregation errors. Detailed commuting data at the census block level, however, are not widely available for most cities in the United States. Different unit scales and unit zone definitions cause inconsistency in analysis results, particularly common in comparison analysis over time, and thus lead to the well-known Modifiable Areal Unit Problem (MAUP) (Niedzielski et al., 2013). Some recent studies have shown great promise in using Global Positioning System (GPS) data and activity travel surveys of individual trip makers in commuting studies (Shen et al., 2013; Kwan and Kotsev, 2015). More accessible and accurate measures of commuting lengths remain very much needed.

## 2.2 EXPLAINING COMMUTING BY SPATIAL FACTORS

There has been a sustained interest in (and debate of) explaining intra-urban variations of commuting lengths or patterns by the land use layout (Giuliano and Small, 1993; Wang, 2000; Sultana, 2002; Horner, 2004, 2007; Horner and Schleith, 2012; Hu and Wang, 2016). In essence, commuting is a spatial process for a worker to overcome the spatial barrier from his/her home to workplace. Therefore, explanation of commuting patterns naturally begins with a focus on the spatial separation of resident (population) and employment locations. Past attempts include modeling how far a residential location is from a job concentration area such as the CBD or from the overall job market, or measuring the need for commuting beyond a local area that is captured by the so-called jobs–housing balance ratio. For example, the jobs–housing imbalance is found to be related to longer commutes and worsened traffic congestion. An area would be considered jobs–housing imbalanced when the number of resident workers differ considerably from the number of jobs (Giuliano and Small, 1993). Cervero (1989), based on data in the San Francisco Bay area and other U.S. cities, found that areas with a severe jobs–housing imbalance were associated with longer commutes. Similarly, based on the 1990 Census Transportation Planning Package (CTPP) data, Sultana (2002) found that the jobs–housing imbalance was a significant determinant of long commutes in the Atlanta metropolitan area. Horner (2002) studied 26 U.S. cities and reached the conclusion that commuting lengths were well-correlated with the jobs–housing balance. Wang (2000) took a step further by comparing the impacts of a jobs–housing imbalance between the commuting distance and time and found that the commuting distance was more sensitive to the jobs–housing imbalance than time.

A few recent studies examined the temporal change in commuting patterns and connected it with land use patterns. For example, Horner (2007) explained the spatial–temporal pattern of intra-urban commuting (mileage and multiple commuting efficiency metrics) from the jobs–housing balance perspective in Tallahassee, Florida from 1990 to 2000. Similarly, Chen et al. (2010) investigated the change in commuting patterns from analyzing the residential and employment distributions in central Texas between 1990 and 2000.

Policies advocating some land use metrics such as the jobs–housing balance differ from other attempts on commuting in that they are explicitly focused on the spatial perspective of commuting, and this lays such approaches open to criticisms, however (Horner, 2004). For instance, Giuliano and Small (1993) found that the relationship between commuting times and the jobs–housing balance ratio was weak in the Los Angeles area. Peng (1997) concluded that such a relationship between commuting and the jobs–housing balance was significant only in cases where jobs and housing were extremely unbalanced. Similarly, Cervero (1996) supported the above finding but defined it more specifically that only in places where job creation was far more than housing production did commute lengths noticeably change. Levinson (1998) also investigated such plausible relationship and found moderate support in Washington, DC. Specifically, he argued that personal choice would play a more significant role in affecting housing/employment decisions than the jobs–housing balance, and therefore, policies aimed at improving the jobs–housing balance in order to alleviate traffic congestion would not perform well.

In the meantime, some studies have been interested in the relationship between job decentralization and commuting. As discovered by Gordon et al. (1991), the average commuting time in the 20 largest metropolitan areas was found to remain stable over time, albeit there was an increase in traffic congestion in cities. Based on the trade-off assumption in the theoretical urban economic model, a more decentralized city would enable more workers to move near employment sub-centers to minimize their commuting costs and thus give rise to shorter commuting time (Gordon et al., 1989a). This is consistent with Downs (1992), who argued that the decentralization of suburban job sites would decrease commuting lengths. To better explain the above commuting paradox, a colocation theory was proposed, which states that workers would change their residence or jobs in order to adapt to the worsening traffic congestion (Gordon et al., 1991). Levinson and Kumar (1994) supported this theory based on their findings of stable commuting time in the Washington metropolitan region and further defined those workers as rational locators. Empirical evidence to this pattern is also found in Crane and Chatman (2003) and Sultana and Weber (2007) who argue that job decentralization would encourage shorter commuting distances.

However, some studies contradict the colocation theory. For example, Cervero and Wu (1998) found that the rapid job decentralization in the San Francisco Bay area, on the contrary, led to a rise in the average commuting distance and time between 1980 and 1990. This opposing finding was also detected in Levinson and Wu's (2005) study. In addition, they also found that the commuting time could vary among different metropolitan structures.

Aguilera (2005) also discovered the same changing pattern in the three largest French metropolitan areas and thus disputed the colocation theory.

## 2.3 EXPLAINING COMMUTING BY NONSPATIAL FACTORS

There are also nonspatial factors that lead to longer commutes beyond what can be predicted by the land use pattern. Some research studies have tried to understand commuting based on the classic urban economic model (Muth, 1969; Mills, 1972), which suggests that the trade-off between commuting and housing dictates a household's residential choice. Specifically, it assumes that workers will move farther away from the CBD and thus commute longer to trade for better housing. Zax and Kain (1991) investigated the impact of long commute distances on quit and move behavior based on the residential choice theory. They found that long commutes resulted from the trade-off between commuting cost and housing consumption encouraging quits and discouraging moves in metropolitan areas with negative wage and housing price gradients. Dubin (1991) found that workers preferred less commuting in terms of time and used suburbanization to shorten their commute times, which was consistent with the prediction of the monocentric model that job decentralization reduces total commuting lengths.

Arguably, it becomes increasingly difficult to apply this simplistic urban model to current urban structures, particularly large metropolitan areas where job decentralization becomes a norm (Horner, 2004). This assumption regarding income and housing may still be applicable to urban areas where the CBD dominates the job market, however. Given the same income level, workers living farther away from the CBD, the largest job center in a city, would need to commute longer, but they would be compensated by more spacious housing. Over time, the trade-off leads to a decentralized housing development and job distribution pattern. Such decentralized patterns are especially common in the United States, where more people are enjoying increased mobility and are capable of living where they want to, without compromising their opportunity to engage in life activities (Horner, 2004). In fact, income is reported to be the other major factor that affects residential segregation besides race–ethnicity in the United States (Massey et al., 2009; Niedzielski et al., 2015). For instance, the number of high-poverty neighborhoods is increasing, and are spreading from central cities to suburban areas (Cooke and Marchant, 2006; Kneebone and Garr, 2010). In addition, structures of jobs are also changing; for example, the number of low-skilled jobs (usually less payed) is growing in many suburban locations and across metropolitan areas in the United States (Niedzielski et al., 2015).

As residents differ a great deal in terms of income, how does income influence one's tolerance to commute and desire of housing space? For example, low-wage workers, on average, are reported to spend a much higher proportion of their income on commuting (6.1 percent) than other workers (3.8 percent); furthermore, the working poor who rent, spend a greater portion of their income on combined costs of commuting and housing (32.4 percent) than other workers (19.7 percent) (Roberto, 2008). Low-wage workers also have significantly lower vehicle ownership than others (Lowe and Marmol, 2013) and thus are more likely to use slower transportation modes such as public transit, carpooling, biking, and walking than their high-wage counterparts (Ross and Svajlenka, 2012; McKenzie, 2014). To this end, higher-wage workers may generally commute longer than lower-wage workers because of their better mobility and economic conditions. Lack of access to data of individual commuters, particularly quality data over a long time period, prevents us from addressing the question directly. Nevertheless, an analysis of commuting variability by neighborhood income levels may still shed light on the issue (Wang, 2003; Hu et al., 2017). Gordon et al. (1989c) investigated the relationship between neighborhood income and commuting and argued that commuting times would increase with income. Rosenbloom and Burns (1993) discovered the same pattern between commuting and income; as income increases, commuting distances would increase for both male and female workers. Wang (2003) measured commuting lengths (in both distance and time) for different wage groups in Cleveland and found that, compared to time, the commuting distance was more sensitive to wage in a way that wealthier neighborhoods commuted more than poorer ones, but the wealthiest neighborhoods shortened their commutes slightly. A recent study by Horner and Schleith (2012) investigated the neighborhood commuting pattern by three income groups (i.e., low, medium, and high) based on their monthly income and found that the average commutes became lengthier as the income increased. More detailed income subgroups may help detect more subtle patterns in terms of income and commuting length, however.

There are also other nonspatial factors that have been commonly investigated in the literature besides income. For instance, people of a racial–ethnic minority (by extension any disadvantaged groups) may commute more than their white counterparts. This is referred to as the "spatial mismatch" (Kain, 1968), as a result of residential segregation, economic restructuring, and the suburbanization of employment. A multiple-worker household is reported to commute longer as a result of the difficulty of coordinating multiple commute trips. Gera (1979) found that older people tend to commute longer, indicating the impact of age on an individual's commuting length. Gordon et al. (1989b) recognized that women have significantly shorter commute lengths than men. Gordon et al. (1989c) suspected that commuting could be related with occupation; and a higher share of industrial employment would offer more opportunities for shorter commuting time. They also examined transportation modes and found that carpooling resulted in longer commuting time than driving alone. Wang (2001) considered workers' socio-demographic characteristics such as race, gender, homeownership status, and education in his model and found significant relationships between some variables (e.g., race and gender) and commuting lengths. Sultana and Weber

(2007) also showed the promise of using workers' socioeconomic characteristics to explain commuting patterns. In short, as discussed above, commuting may be explained by where commuters are and by who they are (Wang, 2003).

## 2.4 MEASURING WASTEFUL COMMUTING

Wasteful or excess commuting is another line of research closely related to the paradigm of interrelatedness between land use and commuting and reflects the overall commuting efficiency in a city. It is measured as the proportion of the actual commute that is over the minimum (optimal or required) commute when assuming that people could freely swap their homes and jobs in a city (Hamilton, 1982; White, 1988; Horner and Murray, 2002). Instead of focusing on the variation of commuting across areas, it highlights how much the overall commuting could be reduced based on the above assumption. In other words, the concept captures the potential (or lack of potential) for a city to optimize commuting without altering the existing land use, and, to some extent, reflects efficiency in its land use layout. Based on the definition of the optimal commute, other commuting efficiency metrics were then designed. Horner (2002) proposed the theoretical maximum commute to represent the most inefficient or costly regional commuting flow pattern given the existing spatial arrangement of workers and jobs. He then defined a metric commuting range as the difference between the theoretical maximum and minimum commute to reveal how much commuting flexibility or capacity is available for commuting. Another metric, the capacity consumed statistic, is expressed as the difference between the actual and optimal commute, divided by the commuting range. It indicates how much of a region's commuting potential (or capacity) has been utilized (accounted for by the observed commute). Both metrics above provide alternative views of a region's commuting efficiency in comparison to wasteful commuting.

Besides the minimum and maximum commute, a theoretical random commute was then put forward (Hamilton, 1982; Charron,

2007; Yang and Ferreira, 2008; Layman and Horner, 2010; Murphy and Killen, 2011), sitting between the lowest and highest theoretical commuting cost scenarios. Alternatively, it represents a commute pattern in which workers are not sensitive to commuting costs in making location decisions. To put it another way, it corresponds to a spatial interaction model with no distance decay effect, that is, no impact of travel costs (Niedzielski et al., 2013). Based upon the theoretical random commute, another policy-relevant commuting efficiency metric named commuting economy is provided. Similar to the capacity consumed statistic, commuting economy also tackles the association between commuting potential and practical utilization. Carried out differently, the theoretical random commute is considered a more realistic representation of the commuting upper bound than the maximum commute in a region.

Though all of the above metrics offer insight into a region's commuting efficiency, this book is specifically interested in wasteful commuting that tackles the connection between the theoretical minimum commute and actual commute because of its fundamental importance in the literature of commuting efficiency, its implication related to the journey to work pattern, land use, and the spatial separation of home and jobs, its policy-related role in benchmarking urban travel efficiency (Horner, 2004; Horner and Schleith, 2012), and, most importantly, its methodological issues that remain unresolved (Horner and Murray, 2002; Fan et al., 2011; Niedzielski et al., 2013).

The concept of wasteful commuting was first proposed by Hamilton (1982) to examine if the classical urban economic model would perform a good job in predicting the mean commute length in a city. Hamilton assumed that both population and employment densities decline exponentially with distance from the city center, and the latter have a steeper gradient than the former. Assuming that residents could freely swap houses, the optimal (minimal) commuting pattern is that people always commute toward the city center and the trips end at the nearest jobs. As a result, the average minimal commute is the difference in average distances of population and employment from the city center. Surprisingly, he found that the actual commute was about 87 percent in excess in comparison to the optimal in 14 U.S. cities. White (1988) argued that the urban commute optimization should be constrained to the existing spatial distribution of homes and jobs and the road network, and she proposed a simple Linear Programming (LP) approach to measure the optimal commute. Surprisingly, White's model returned only 11 percent wasteful commuting for the same study areas used by Hamilton. The large gap in the results by Hamilton and White has led to a sustained debate on how to accurately measure wasteful commuting and generated a large volume of literature from multiple disciplines.

Some attributed the discrepancy to the scale (zonal) effect. Hamilton (1989) cautioned researchers of assuming an optimal intrazonal commute and pointed out that a larger zone size might lead to less wasteful commuting. In practice, the LP result usually yields a high proportion of optimal commute trips within a zone; for example, 90.7 percent of optimal commute trips in Small and Song (1992) were intrazonal commute. If one adopts average reported intrazonal commuting time for all zones in a study area, the optimal commute is likely to be overestimated. An inflated optimal commute leads to an underestimated wasteful commuting such as the case for White (1988) that was only 11 percent. Based on a smaller zone, that is, Traffic Analysis Zones (TAZs), Small and Song (1992) implemented White's LP approach in Los Angeles and found 66 percent was wasteful commuting, which was substantially higher than White's but lower than Hamilton's. Their finding confirmed Hamilton's proposition of scale effect. In addition, they suggested a normatively neutral term excess commuting instead of wasteful commuting. Horner and Murray (2002) linked the issue to the MAUP, which is well known to geographers (Openshaw and Taylor, 1979). They further validated the impact of spatial unit definition on the estimation of wasteful commuting and suggested using zonal data as disaggregate as possible.

Others suspected that different metrics of commuting length might play a role (Hamilton, 1989). Hamilton (1982) used distance while White (1988) used travel time. Most found a high consistency between the two and rejected that it was a major factor causing the discrepancy (e.g., Cropper and Gordon, 1991; Small and Song, 1992).

Some believed that job decentralization might account for most or all of the wasteful commuting (Merriman et al., 1995; Suh, 1990). However, as argued by Giuliano and Small (1993), the direction of job decentralization's impact on commuting could be ambiguous. On the one side, it may encourage urban sprawl, reduce the land use intensity and thus increase commute lengths. On the other side, it may also improve the jobs–housing balance in many areas, particularly suburbia, and alleviate the need of lengthy commutes to downtown. Following the suggestions of Hamilton (1982), several studies searched for factors beyond land use that prevented people from attaining the optimal commute. Some emphasized the variability of the labor participation rate across households as it is harder for households with multiple workers to optimize commuting for individuals (Kim, 1995; Thurston and Yezer, 1991). Certainly, residential choice is hardly made solely for the purpose of minimal commuting and often involves a complex decision considering also housing and neighborhood attributes (Cropper and Gordon, 1991). Not all workers are mobile in terms of residential choices and thus limit their likelihood of relocating to save commute (Buliung and Kanaroglou, 2002). Furthermore, the boundary effect for a study area may also affect the magnitude of wasteful commuting as Frost et al. (1998) found in their study of British cities. Our fascination of wasteful commuting is also linked to its implication in public policy (Fan et al., 2011). Many have studied the issue for various transportation or land use related policy scenarios that could reduce commuting costs and their related environmental and economic impacts (Boussauw et al., 2012; Horner, 2002; Ma and Banister, 2006; O'Kelly and Lee, 2005; Rodríguez, 2004; Scott et al., 1997; Yang, 2008).

This book directs the attention back to the measurement of wasteful commuting given a land use pattern (i.e., locations of resident workers and employment). Is it possible to design a study as disaggregate as possible to mitigate (or perhaps be even free from) the scale effect for estimating the required commute, as suggested by Horner and Murray (2002)? Given the concern related to privacy, it is unlikely for researchers to access a large-scale commuting dataset of individual households geocoded to their home addresses and workplaces. Our approach is to simulate individual employment and resident worker locations and the trips between them in order to alleviate the concern of unreliable estimates of intrazonal commute and also to improve the accuracy of estimating interzonal commute lengths.

Another issue is how an actual commute is measured. Some commuting literature (though not directly related to wasteful commuting) suggests the unreliability of measuring commute length by travel time. It could be misleading as travel time by slower modes such as carpooling, public transit, biking, and walking is much longer than driving alone for the same distance. The journey-to-work survey data such as the CTPP often contain some erroneous records (e.g., commute trips of several hours for traveling only a few miles; for example, the reported mean travel time from tract 36.04 to tract 11.04 in Baton Rouge, Louisiana was 3.7 hours [with an estimated travel distance 5.4 miles], 2.8 hours from tract 22 to 26.01 [with an estimated travel distance 2.6 miles], and so on). There is also a concern for inconsistency in the way respondents report commute times (e.g., whether including "mental time" as noted in Wang 2003). Wasteful commuting calculated at different times may also vary (Frost et al., 1998; Yang, 2008). As wasteful commuting is the difference between actual and optimal commuting, it is an unfair comparison to define actual commuting by reported time and optimal commuting by estimation. This book proposes to measure both by estimated travel time and distance through the road network and therefore identify the true extent of wasteful commuting.

# CHAPTER 3

# Study Area and Data

T HIS CHAPTER INTRODUCES THE study area of case studies used in the book and corresponding data sources. Section 3.1 briefly describes the study area. Section 3.2 presents the data sources used in case studies throughout the book. For your convenience, we provide major datasets and program files in the data folder along with the book (downloadable via the links provided in the Preface). They are listed in Section 3.3.

## 3.1 STUDY AREA

As the core of the Baton Rouge Metropolitan Area, East Baton Rouge Parish in Louisiana is selected as the study area throughout the book. Note that a parish in Louisiana corresponds to a county in other states in the United States. This parish has an area of 471 square miles including the City of Baton Rouge (the state capital, one of the fastest growing areas in the South) in the middle, Baker and Zachary in the north. It is surrounded by eight parishes. There are two major rivers in this region—the Mississippi River and Amite River—forming the natural borders of the study area. The Mississippi River on the west divides East Baton Rouge Parish from West Baton Rouge Parish and Pointe Coupee Parish, while the Amite River on the east segregates this

Parish from Livingston Parish and St. Helena Parish. In terms of land use, these neighboring parishes are mostly composed of rural areas (Antipova et al., 2011). The northeastern, northwestern, and the most southern parts of East Baton Rouge Parish are also recognized as rural areas that may serve as the buffer zone between the urbanized areas (e.g., the most central area that has high road density) in this parish and other neighboring parishes. Interested readers may refer to Ikram et al. (2015) for the map and more description about this region. Given the above distribution patterns, edge effects, a common phenomenon that could result in unreliable conclusions due to the incomplete consideration of the impacts of border areas, might be of little influence in this study area (Wang et al., 2011). For simplicity, hereafter, the study area is referred to as Baton Rouge. Figure 3.1 illustrates the study area overlaid with census tract boundaries and job density patterns.

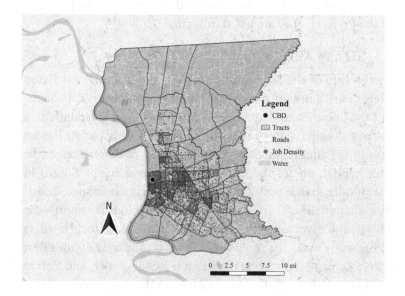

FIGURE 3.1   Baton Rouge region of 2010.

## 3.2 DATA SOURCES

The major data source used throughout the book is three Census for Transportation Planning Packages (CTPPs): the 1990 and 2000 CTPPs from the Bureau of Transportation Statistics (2014), and the most recent 2006–2010 CTPP from the American Association of State Highway and Transportation Officials (2014). Note that the 1990 or 2000 CTPP was extracted from the long form–decennial census, and the 2006–2010 CTPP was based on the 5-year American Community Survey (ACS) 2006–2010 (hereafter simply referred to as 2010). All CTPPs consist of three parts: (1) part 1 on information by place of residence such as the number of resident workers in a zone; (2) part 2 on records by place of work such as number of jobs in a zone; and (3) part 3 on journey-to-work flow such as number of commuters from a residence zone to a workplace zone and the average commuting time breakdowns of different transportation modes including driving alone, carpooling, public transit, biking, and walking.

The CTPP data are usually provided at multiple levels of aggregations (zones) such as traffic analysis zones (TAZ), census tracts, states, and metropolitan statistical areas (MSA). However, there is inconsistency in the area unit used across the three CTPP data for Baton Rouge: TAZ in 1990, multiple zonal levels in 2000 (census tract, census block group, and TAZ for parts 1 and 2, only census tract for part 3), and census tract and TAZ in 2006–2010. We chose census tracts as the unit to use throughout. Luckily in Baton Rouge, the 1990 TAZs were mostly components of the census tract for easy aggregation with only very few minor exceptions. There were 85, 89, and 91 census tracts in Baton Rouge in 1990, 2000, and 2010, respectively (after excluding the 2010 airport tract where no records of any resident workers or jobs are provided in the data). The slightly increased number of tracts in later years was simply the result of split tracts from earlier years (i.e., 2000 vs. 1990, 2010 vs. 2000). This enabled us to integrate the data in three time epochs based on the 85 census tracts in 1990 when needed. Corresponding spatial datasets in GIS including census tracts, TAZs, and road networks are extracted from the Topologically Integrated Geographic Encoding and Referencing (TIGER) Products, 1994, 2000, and 2010 from the U.S. Census Bureau. We are aware of the time gaps of using the 1994 and 2010 GIS data to match the 1990 and 2006–2010 CTPP, respectively; they are the best data accessible to us, however.

The National Land Cover Database (NLCD, http://www.mrlc. gov), a national land cover product created by the Multi-Resolution Land Characteristics (MRLC) Consortium, is used to improve the accuracy of individual trip simulation, in particular, the simulation of trip destinations (see Section 4.3 in Chapter 4). Three NLCD products—NLCD 1992, 2001, and 2011—are employed to match with the above CTPP and TIGER data. The NLCD has a spatial resolution of 30 square meters, and only one land cover type is recorded in each pixel, a 30 m × 30 m polygon. It provides a uniform land cover classification across the entire United States and is perhaps the most accessible and commonly used national land cover map (Jin et al., 2013). In this book, the high intensity developed areas that are commonly interpreted as commercial/industrial lands are used to define the geographic areas for simulated job locations. The geographic areas of resident workers, however, are calibrated on the basis of census block population data obtained from the U.S. Census Bureau due to its better accuracy in capturing the residential pattern than the NLCD. See Table 3.1 for a summary of all data sources used in this book.

## 3.3 LIST OF PROVIDED DATA

The following datasets, which are the inputs of case studies in the book, are prepared and provided under the folder `data`:

1. A geodatabase `br2010.gdb`. It includes (a) a polygon feature class `trt` for 91 census tracts (again, the airport tract has no records of resident workers or jobs provided in the 2006–2010 CTPP, and thus is excluded from the analyses); (b) the

TABLE 3.1  Summary of All Data Sources

| Data Layer | Year | Spatial Scale | Format | Source | Purpose |
|---|---|---|---|---|---|
| CTPP | 1990 | TAZ | Text/Excel file | BTS | Total number of workers, jobs, and commuters |
| | 2000 | Census tract | Text/Excel file | BTS | |
| | 2010[a] | Census tract | Text/Excel file | AASHTO | |
| Zone boundary | 1994 | TAZ | Vector/shapefile | TIGER/Line | Define boundary of zone units |
| | 2000 | Census tract | Vector/shapefile | TIGER/Line | |
| | 2010 | Census tract | Vector/shapefile | TIGER/Line | |
| Population | 1990 | Census block | Text/Excel file | Census | Spatial extent of residential areas |
| | 2000 | Census block | Text/Excel file | Census | |
| | 2010 | Census block | Text/Excel file | Census | |
| NLCD | 1992 | 30 m × 30 m cell | Raster/tif file | MRLC | Spatial extent of workplaces |
| | 2001 | 30 m × 30 m cell | Raster/tif file | MRLC | |
| | 2011 | 30 m × 30 m cell | Raster/tif file | MRLC | |
| Road network | 1994 | – | Vector/shapefile | TIGER/Line | Define entire road network |
| | 2000 | – | Vector/shapefile | TIGER/Line | |
| | 2010 | – | Vector/shapefile | TIGER/Line | |

[a] Based on the 2006–2010 CTPP, 2010 is used here and in all other tables for simplicity.

corresponding point feature class trtpt for the 91 census tracts,* containing fields TotWrk and TotJob for the number of resident workers and jobs extracted from CTPP part 1 and part 2, respectively; (c) a point feature class BRCenter for the central business district (CBD) of the study area, Baton Rouge, Louisiana; (d) a table ctppTT with 8,281 records (91 * 91) that contains a field TT _ AllMeans representing the average all-travel-mode commute times in minutes between every two tracts (extracted from CTPP part 3) and a field O _ TotWrk saving the number of resident workers in each residential tract; (e) a table ctppFlow with 8,281 records (91 * 91) containing the observed number of commuters between every two tracts (extracted from CTPP part 3); and (f) a feature dataset rd _ 2010 containing a network dataset rd _ 2010 _ ND and all feature classes associated with rd _ 2010 _ ND including the edges rd _ 2010 and junctions rd _ 2010 _ ND _ Junctions.

2. A text file wrk _ job _ inBR _ 2010.txt containing three fields (a) ID—the trtID _ new of each census tract, (b) Res—the number of resident workers living in each tract, and (c) Emp—the number of jobs located in each tract. This file is used as an input to the Linear Programming (LP) algorithm that will be introduced in Section 6.2 and Section 6.3 in Chapter 6. As the calculation of wasteful commuting requires an equal number of worker and job counts, the Res and Emp were extracted from CTPP part 3 commuter flow data and represent the number of resident workers who both live and work in Baton Rouge and the number of employment where both its location and the employee's home address are within Baton Rouge, respectively. They are different from TotWrk and TotJob in trtpt described above.

---

* A field "trtID _ new" is added to the attribute tables of both trt and trtpt to assign a unique index value (1–91) to the tracts and their centroids.

The following program files that implement the proposed methods in the book are provided under the data folder as well.

3. An executable program TSME.exe with four modules for tasks in traffic simulation.* Specifically, Module 2 and Module 3, respectively, are used for simulating resident workers and jobs in Section 4.3 and O-D trips in Section 4.4 of Chapter 4.

4. Two R program files—WasteComm _ LP.R that implements the LP approach in Section 6.3 and WasteComm _ ILP.R that implements the proposed Integer Linear Programming (ILP) approach in Section 6.4 in Chapter 6.

5. A toolkit "Features To Text File.tbx" and the corresponding folder Scripts include a deprecated ArcGIS tool extracted from the "Samples Toolbox" (http://desktop. arcgis.com/en/arcmap/10.3/tools/samples-toolbox/write-features-to-text-file.htm). It will be used in simulating resident workers and jobs in Section 4.3 in Chapter 4.

For convenience, to the following major datasets that are to be produced in case studies in the book are provided under the data folder as well. However, it is not anticipated that one can obtain exactly the same results (e.g., mean commuting distance of 6.25 miles in Baton Rouge in 2010 described in Section 5.1 in Chapter 5) by following the methods detailed in each case study based on the provided data listed here. This is due to the randomness nature of the proposed simulation methods. More detail about the methods will be explained in Chapter 4.

6. A text file wrk _ job _ 2010.txt for the number of resident workers and employment in census tracts extracted

---

* Depending on the operating system, one may be asked to download and install the .Net framework.

from CTPP part 1 and part 2, respectively, and other more fields about the attributes of a tract. It should be noted that this file is different from wrk_job_inBR_2010.txt that was mentioned above in (2). It was created simply by exporting from the attribute table of trtpt. It will be created and used in the simulation process in Section 4.3 in Chapter 4.

7. Two text files oddist.txt and odtime.txt for the network distances and travel times (adjusted by incorporating intrazonal distances and times) between census tracts, respectively. They will be created and used in measuring wasteful commuting using the LP approach in Section 6.3 in Chapter 6.

8. A text file trip_mtx.txt for the observed number of interzonal commuting trips extracted from CTPP part 3. It is created by exporting the attribute table of ctppFlow we mentioned in (1). It will be used to simulate the locations of trip origins and destinations in Section 4.3 in Chapter 4.

9. A text file trt.txt for the spatial extent of the census tract boundaries in 2010. It will be generated and used in the simulation process in Section 4.3 in Chapter 4.

10. Two text files simu_Os.txt and simu_Ds.txt for 500,000 simulated trip origins and destinations, respectively. They will be generated in the simulation process in Section 4.3 in Chapter 4.

11. A text file simu_trips.txt for 75,000 simulated individual trips. It is to be created in the steps in Section 4.4 of Chapter 4.

# Monte Carlo Simulation Method

A S STATED IN SECTION 2.1 in Chapter 2, most existing methods of measuring commuting lengths/patterns are affected
by the aggregation errors and scale effect. To obtain more accurate
and reliable estimates, we propose simulation-based approaches
in this book. Specifically, two Monte Carlo simulation methods
were developed—one is applied to simulate locations of individual resident workers and jobs that are consistent with their spatial
distributions across the area unit like census tracts, and the other
to simulate individual trips that are proportional to the existing
journey-to-work trip flows. By doing so, aggregated trips extracted
from survey data such as the Census for Transportation Planning
Packages (CTPP) are disaggregated into the individual (point)
level. This can help mitigate the aggregation errors and scale effect
and hence permit more accurate estimations of commute lengths
in both analyses of commuting patterns and commuting efficiency
(wasteful commuting). Section 4.1 provides a brief introduction to
the Monte Carlo simulation technique. Section 4.2 presents the
common applications of Monte Carlo simulation in geographic

studies. Section 4.3 and Section 4.4 discuss the principles of two Monte Carlo simulation methods—one in simulating locations of trips ends, that is, resident workers and jobs; and the other in simulating trip distribution, that is, trips between resident workers and jobs. Step-by-step instructions of the technical details to implement both methods are provided in each section.

## 4.1 INTRODUCTION TO MONTE CARLO SIMULATION

The Monte Carlo method generates suitable random numbers of parameters or inputs to explore the behavior of a complex system or process. The random numbers generated follow a certain probability distribution function (PDF) that describes the occurrence probability of an event. Some common PDFs include:

- A *normal* distribution is defined by a mean and a standard deviation. Values in the middle near the mean are most likely to occur, and the probability declines symmetrically from the mean.

- If the logarithm of a random variable is normally distributed, the variable's distribution is *lognormal*. For a lognormal distribution, the variable takes only positive real values, and may be considered as the multiplicative product of several independent random variables that are positive. The left tail of a lognormal distribution is short and steeper and approaches toward 0, and its right tail is long and flatter and approaches toward infinity.

- In a *uniform* distribution, all values have an equal chance of occurring.

- A *discrete* distribution is composed of specific values, each of which occurs with a corresponding likelihood. For example, there are several turning choices at an intersection, and a field survey suggests a distribution of 20 percent turning left, 30 percent turning right, and 50 percent going straight.

In a Monte Carlo simulation, each set of random samples is called an iteration and is recorded. This process is repeated a large number of times. The larger the number of iteration times is simulated, the better the simulated samples conform to the predefined PDF. Therefore, the power of a Monte Carlo simulation relies on a large number of simulations. By doing so, the Monte Carlo simulation provides a comprehensive view of what may happen and the probability associated with each scenario.

Clearly, the key to a Monte Carlo simulation is the generation of reliable random numbers following a predefined PDF. Many software packages provide various random number generators corresponding to the common PDFs discussed above such as Matlab and R. For a custom PDF not provided in the software, one can use the inverse transformation method to generate sample numbers at random from a probability distribution defined by its cumulative distribution function (CDF). Hence, this method is also known as inverse CDF transformation. Specifically, one can generate a continuous random variable $X$ by (1) generating a uniform random variable $U$ within (0, 1), and (2) setting $X = F^{-1}(U)$ for transformation to solve $X$ in terms of $U$. Similarly, random numbers following any other probability distributions could be obtained.

To illustrate the process of a Monte Carlo simulation, we present a simple example of simulating traffic at an intersection using Excel. Table 4.1 lists the observed traffic data at an intersection.

Suppose one car is approaching and about to move across the intersection. Let $U$ denote the random number generated by the RAND() function in Excel, a generator to return a random number

TABLE 4.1    Observed Traffic at an Intersection

| Moving Direction | Vehicles/ Hour | Relative Frequency | Cumulative Frequency |
|---|---|---|---|
| Go straight | 525 | 0.525 | 0.525 |
| Turn left | 223 | 0.223 | 0.748 |
| Turn right | 252 | 0.252 | 1.000 |
| Total | 1,000 | 1.000 | |

TABLE 4.2    Results of the First Five Simulations

| Car ID | Random Number $U$ | Simulated Moving Direction |
|--------|-------------------|----------------------------|
| 1 | 0.363792442 | Go straight |
| 2 | 0.244797218 | Go straight |
| 3 | 0.728901323 | Turn left |
| 4 | 0.822877317 | Turn right |
| 5 | 0.151663398 | Go straight |

following a uniform distribution within the range of [0, 1]. The car's movement pattern follows the observed frequency distribution:

- If $0 \leq U < 0.525$, then the car goes straight;

- If $0.525 \leq U < 0.748$, then the car turns left;

- If $0.748 \leq U < 1.000$, then the car turns right.

The above process was repeated 2,000 times in Excel to simulate the movement patterns of 2,000 cars at the intersection. Table 4.2 shows the first five simulation results. The simulations report that about 51.85 percent of the 2,000 cars go straight, 22.40 percent turn left, and 25.75 percent turn right. The probabilities are quite close to the observed ones shown in Table 4.1. In general, the larger the simulation sample is, the more reliable the results are.

## 4.2 MONTE CARLO SIMULATION IN SPATIAL ANALYSIS

The Monte Carlo simulation technique is widely used in spatial analysis. Here, we briefly discuss its applications in spatial data disaggregation and statistical testing.

Spatial data often come as aggregated data in various area units (sometimes large areas) for various reasons. One likely cause is the concern of geo-privacy. Others include administrative convenience, integration of various data sources, and limited data storage space among others. Several problems are associated with the analysis of aggregated data such as

- Modifiable areal unit problem (MAUP), i.e., instability of research results when data of different area units are used.

- Ecological fallacy, when one attempts to infer individual behavior from data of aggregate area units (Robinson, 1950).

- Loss of spatial accuracy when representing areas by their centroids in distance measures.

Therefore, it is desirable to disaggregate data in area units to individual points in some studies. For example, Watanatada and Ben-Akiva (1979) used the Monte Carlo technique to simulate representative individuals distributed in an urban area in order to estimate travel demand for policy analysis. Wegener (1985) designed a Monte Carlo–based housing market model to analyze location decisions of industry and households, corresponding migration and travel patterns, and related public programs and policies. Poulter (1998) employed the method to assess uncertainty in environmental risk assessment and discussed some policy questions related to this sophisticated technique. Luo et al. (2010) used it to randomly disaggregate cancer cases from the zip code level to census blocks in proportion to the age–race composition of block population and examined the implications of spatial aggregation error in public health research. Gao et al. (2013) used it to simulate trips proportionally to mobile phone Erlang values and to predict traffic-flow distributions by accounting for the distance decay rule.

Another popular application of the Monte Carlo simulation in spatial analysis is to test statistical hypotheses using randomization tests. The tradition can be traced back to the seminal work by Fisher (1935). In essence, it returns test statistics by comparing observed data to random samples that are generated under a hypothesis being studied, and the size of the random samples depends on the significance level chosen for the test. A major advantage of Monte Carlo testing is that investigators could use flexible informative statistics rather than a fixed, known distribution theory. Besag and Diggle (1977) described some simple Monte Carlo significance tests in the analysis of spatial data including point patterns, pattern similarity, and space–time interaction. Clifford et al. (1989) used a Monte Carlo simulation technique to assess statistical tests for the correlation coefficient or the covariance between two spatial processes in a spatial autocorrelation. Anselin (1995) used Monte Carlo randomization in the design of statistical significance tests for global and local spatial autocorrelation indices. Shi (2009) proposed a Monte Carlo–based approach to test whether the spatial pattern of actual cancer incidences is statistically significant by computing p-values based on hundreds of randomized cancer distribution patterns. Wang et al. (2017) developed a local indicator of a colocation quotient with a statistical significance test. The local indicator can be used to detect the spatial correlation patterns between types of point incidents, and the significance test—designed based on the Monte Carlo simulation—can determine the statistical significance level for the indicator estimates. Most recently, Hu et al. (2018) designed a spatio-temporal kernel density estimation (STKDE) approach to predicting crime hotspots. To examine the statistical significance of detected hotspots, they generated multiple spatio-temporal random sample events using the Monte Carlo simulation method and then compared them with the observed pattern.

The next two sections introduce a specific application of the Monte Carlo simulation in data disaggregation, one from area-based aggregated data to individual points (e.g., from area-based job count data to individual job points) and another from area-based flow data to individual OD trips (e.g., from area-based commuting flow count data to individual journey-to-work trips). Both are commonly encountered in spatial analysis.

## 4.3 MONTE CARLO SIMULATION OF RESIDENT WORKERS AND JOBS

To mitigate the effect of aggregation errors and scale effect commonly seen in commuting studies, we begin with simulating locations of trip ends, that is, a simulation of individual locations of resident workers and jobs that are proportional to their corresponding numbers within each census tract (i.e., following a discrete distribution). Land use patterns are considered for better simulation performance. Specifically, the number of simulated trip origins is proportional to the number of resident workers in a tract, the number of trip destinations is proportional to the number of jobs there, and both are randomly distributed within areas of residential land use (from census block data) and areas of commercial/industrial land use (from the NLCD) in the tract's boundary, respectively (refer to Hu et al. 2017 for the distribution of commercial/industrial land use patterns in Baton Rouge derived from NLCD 2011). In other words, denoting the total numbers of simulated and actual commuters by $n$ and $N$, respectively, and given the number of resident workers and employment in tract $i$ from the CTPP as $R_i$ and $E_i$, the number of simulated individual workers and jobs in tract $i$ are $(n/N)R_i$ and $(n/N)E_i$, respectively. The value of $n$ is selected by balancing accuracy and computational efficiency (i.e., larger numbers of simulated points and subsequently OD trips improve accuracy but demand more computing power). The following illustrates the implementation process based on the Monte Carlo simulation.

1. *Calculating spatial extent of each census tract.* Based on the spatial extent of the overlapped area of a tract and either the residential (or commercial/industrial) land use (this is recorded in the text file trt.txt by using an ArcGIS tool "Write Features To Text File"), we derive the maximums and minimums for its $X$ and $Y$ coordinates such as $X_{max}$, $X_{min}$, $Y_{max}$, and $Y_{min}$.

2. *Generating X and Y coordinates in corresponding ranges.* Use the Monte Carlo simulation to generate a random number $X_i$ within the range $[X_{min}, X_{max}]$ and another one $Y_i$ within $[Y_{min}, Y_{max}]$ following a uniform distribution for each.

3. *Determining if a point is located within a census tract.* Based on an algorithm built upon Taylor's (1989) ray method,

detect whether a point $(X_i, Y_i)$ is located inside or outside of a census tract's residential (or commercial/industrial) land use areas. If inside, it is retained as a trip origin (or destination); if outside, it is disregarded.

A prototype program, written in C# language, termed *traffic simulation modules for education* (TSME), was developed to automate the above process. The commonly known Urban Transportation Modeling System (UTMS) or Urban Transportation Planning System (UTPS) typically uses the *four-step travel demand model*, composed of trip generation, trip distribution, mode choice, and trip assignment. Each step requires significant efforts in data collection, model calibration, and validation. Even with advanced transportation modeling packages, implementation of the whole process is often arduous or infeasible. With minimal data requirements (using publicly accessible data of road networks and land use), TSME was developed to illustrate the major steps of travel demand forecasting (excluding the step of "mode choice") in distinctive modules. It is composed of four modules, roughly in correspondence to the four-step travel demand forecast model except for the absence of the "mode choice" step and the addition of the "interzonal trip estimation" module. Based on the number of resident workers and jobs in a geographic unit such as a census tract, Module 1 uses a gravity model to estimate the zone-level traffic between tracts. Module 2 uses the Monte Carlo method to randomly simulate individual trip origins (proportional to the distribution of resident workers) and destinations (proportional to the distribution of jobs). Again, based on the Monte Carlo method, Module 3 connects the origins and destinations randomly and caps the volume of OD trips between each pair of zones proportionally to the observed interzonal traffic volume (or, the estimated interzonal traffic from Module 1 when no observation data is available). Finally, Module 4 calibrates the shortest routes for all OD trips, measures the simulated traffic at traffic monitoring stations, and compares it to the observed traffic at those stations to validate the model. The modular design in TSME is to allow for some users to use a module for applications beyond traffic simulation. Interested readers can refer to Hu and Wang (2015b) for more detail. A recent example of using the TSME for simulating interzonal commuting trips can be found in Li et al. (2018).

We will use Module 2 (under the tab "Monte Carlo Simulation for O's & D's") in TSME to simulate trip origins and destinations. The following reports the technical details for simulating locations of individual resident workers and jobs in 2010. For illustration purpose, we only show the simulation process without considering land use data. Advanced users may use the intersect analysis in GIS to attain only the residential (or commercial/industrial) areas in a tract and then apply the steps below for improved simulation accuracy.

### Step 1  Exporting Census Tract Boundaries to a Text File in ArcGIS

In ArcCatalog, locate the tool "Write Features To Text File" under the provided toolkit "Features To Text File" and double-click it to open the dialog window. Select the feature class trt for the Input Features, name the Output text File trt.txt, and the default setting "Locale Decimal Point" is ok for Decimal Separator Character. Click OK to execute it.

### Step 2  Exporting Census Tract Resident Worker and Job Counts to a Text File in ArcGIS

Load the feature class trtpt, where the number of resident workers and jobs in each tract are stored, into ArcMap. Right-click to open its attribute table. Then click the Table Options menu on the top-left corner and select Export. Name the Output table as wrk _ job _ 2010.txt (select Text File as the Save as type) and click OK to execute it.

### Step 3  Simulating Origins and Destinations in TSME

On the main menu of the TSME program, select the second tab "Monte Carlo Simulation of O's & D's." As shown in Figure 4.1, (1) import the polygon boundary text file trt.txt prepared in

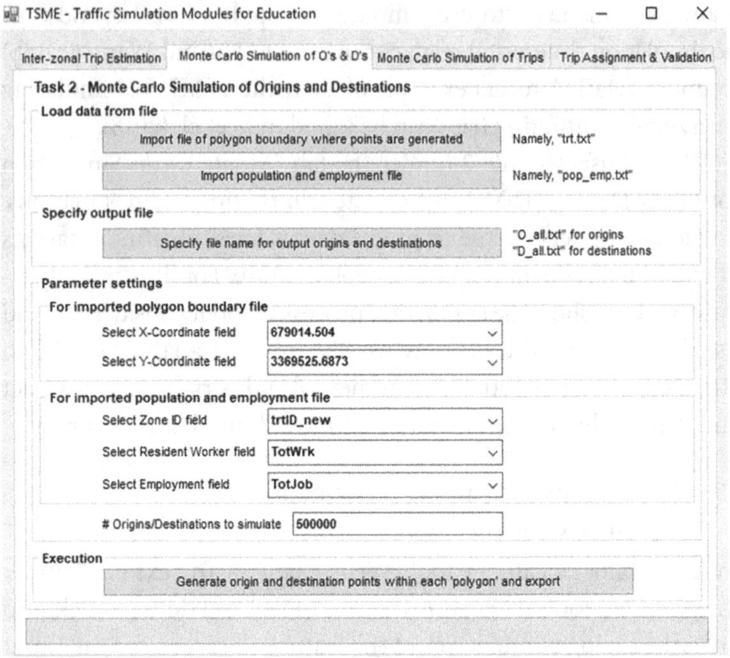

FIGURE 4.1    TSME interface for the Monte Carlo simulation for O's & D's module.

step 1 and import the population and employment file wrk_job_2010.txt; (2) specify the output files for trip origins and destinations (e.g., simu_Os.txt and simu_Ds.txt); (3) under "Parameter settings," select the corresponding fields associated with input files (automatically chosen by the program as shown in Figure 4.1) and set 500,000 as the #Origins/Destinations to simulate. Note that 500,000 is suggested for a balance between computation time and sufficient sample sizes for origins and destinations. Click the button in the Execution panel to run the module.

For an illustration of the process, we selected only four census tracts in Baton Rouge as an example. Given the zonal level spatial patterns of resident workers and jobs in the four selected tracts in Figure 4.2a,c, respectively, Figure 4.2b,d shows corresponding simulated individual locations of resident workers and jobs.

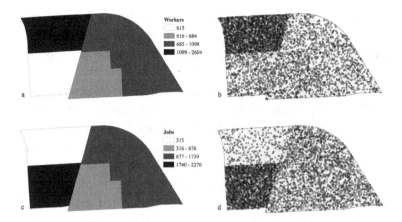

FIGURE 4.2 Spatial distribution of (a) resident workers in zones, (b) simulated resident workers, (c) jobs in zones, and (d) simulated jobs.

Interested readers should refer to Hu et al. (2017) for results after residential and job land uses are accounted for.

A recent study (Wang et al., 2018) on the urban population density pattern in the Chicago Metropolitan area used the aforementioned TSME module to simulate the same amount of population (residents) as reported in the census, that is, 8.2 million. Specifically, the study utilized the 2010 census data at the census block level and the 2010 Land Use Inventory data from the Chicago Metropolitan Agency for Planning (CMAP) Data Hub (https://datahub.cmap.ill inois.gov/group/land-use-inventories) to improve the accuracy of simulation by limiting the simulated individual residents to only "residential blocks." Residential block refers to a whole block or part of a block that was identified as residential land use by overlaying the layers of census blocks and land uses. Once the point layer of simulated residents was prepared, the study proceeded by aggregating them back to various predefined uniform area units such as square, triangle, and hexagon, and deriving what function best captured the corresponding population density patterns. It then compared the results to the best fitting density function based on the census data in area units such as census tracts and block groups. The case study indicated that the best fitting density function was exponential for data in both census units (consistent with the literature), however, the logarithmic function became a better fit when uniform area units were used.

This finding suggested that the commonly observed exponential urban density pattern (Clark, 1951) might be an artifact of census data reported in artificially defined census units. Any studies based on these data are subject to criticisms such as the MAUP, unfairness of sampling (Frankena, 1978), and inaccuracy or uncertainty in distance measure by using centroids to represent areas (i.e., aggregation errors). The simulated data enabled one to aggregate population back to area units of any scale in any shape so that the scale effect and the zonal effect were examined explicitly. When an area was represented as the average location of individuals within the area, it yielded a more accurate centroid to facilitate the distance calibration. In short, the core technique behind the solution to those concerns was the Monte Carlo simulation of individual resident locations.

## 4.4 MONTE CARLO SIMULATION OF COMMUTING TRIPS

The previous section returns simulated points of resident workers and jobs according to their spatial distributions across the area unit such as census tracts. This section shows another application of the Monte Carlo simulation in connecting resident workers and jobs together to form commuting trips, or more generally, disaggregating area-based flow data to individual OD trips.

To achieve this, another simulation process is designed for trip distribution. Specifically, we simulate the trips between individual locations of workers and jobs that are consistent with the actual zonal-level flows. Similarly, as the actual zonal-level flow from a residential worker tract $i$ to an employment tract $j$ (extracted from the CTPP) is $x_{ij}$, the simulated flow when aggregated at the zonal level should be $(n/N)x_{ij}$, that is, proportional to the actual journey-to-work pattern. This is implemented by another Monte

Carlo simulation process by utilizing the previously simulated points of resident workers (O) and jobs (D), for example, tables simu _ Os.txt and simu _ Ds.txt in step 4 in Section 4.3. The following illustrates the principle of this Monte Carlo simulation process.

1. *Randomly choosing an origin point in a zone.* Say a zone $m$ contains $O_m$ origins. Use the Monte Carlo method to randomly choose an origin, denoted as $O_i$, where $i \in [1, O_m]$.

2. *Randomly choosing a destination point in another zone.* Similarly, randomly choose a destination from a zone with $D_n$ destinations, denoted as $D_j$, where $j \in [1, D_n]$.

3. *Forming OD trips and counting frequency.* Link $O_i$ and $D_j$ to form a trip. Cumulate the count for trips from an origin zone to a destination zone, denoted by $F_{mn}$.

4. *Capping the number of OD trips.* Continue the iterations of the above three steps until $F_{mn}$ reaches the predefined zonal-level flow $F_{mn0}$, i.e., $(n/N)x_{ij}$.

The process is automated in Module 3 (under the tab "Monte Carlo Simulation of Trips") in the TSME program interface. There are two options offered by Module 3 to simulate OD trips. The first option is to generate completely random trips by following a uniform distribution, which may be desirable for some users when no prior knowledge of zone-level traffic is available. The other is to pair the origins and destinations that follow a discrete frequency distribution specified by the estimated interzonal trips obtained from Module 1, or the observed trips from CTPP part 3 in our case. For our purpose, we will use the second option.

### Step 4 Simulating Individual OD Trips in TSME

In TSME, select the third tab "Monte Carlo Simulation of Trips." As shown in Figure 4.3, import the files for simulated origins and destinations (simu _ Os.txt and simu _ Ds.txt) derived

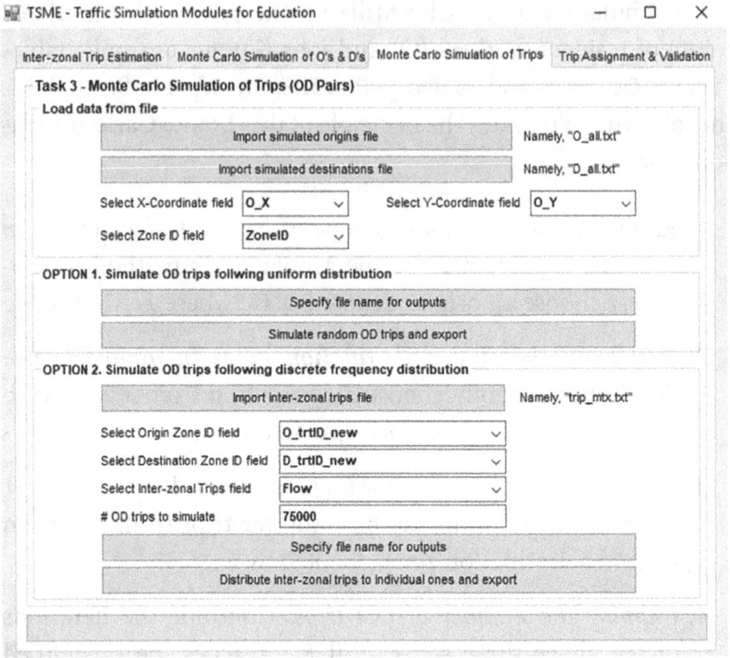

FIGURE 4.3    TSME interface for the Monte Carlo Simulation of Trips module.

from step 3 in Section 4.3 and select corresponding fields for $X$ and $Y$ coordinates and tract ID (again automatically populated by the program as shown in Figure 4.3). For Option 1 that is aimed to generate completely random trips by following a uniform distribution, specify the output file for the OD trips (e.g., OD _ random.txt by default), and click "Simulate random OD trips and export" to execute it. For Option 2 that is designed to generate individual trips following a discrete frequency distribution specified by the observed interzonal trips obtained from CTPP part 3, import the observed interzonal trips file (trip _ mtx.txt) that comes with the book, select corresponding fields (also populated as shown in Figure 4.3), set 75,000 as # OD trips to simulate,* specify the output file for the OD trips (e.g., `simu _ trips. txt`), and click "Distribute interzonal trips to individual ones and export" to execute it.

In the resulting OD trip file `simu _ trips.txt`, the first 75,000 rows are the $X$ and $Y$ coordinates for the 75,000 origin points identified by their unique `Route _ ID`, and the next 75,000 rows are the $X$ and $Y$ coordinates for the corresponding destinations and the same `Route _ ID`.

The proposed simulation approaches disaggregate the reported zonal commuting trips into individual trips and hence permits a more accurate estimation of commute distances by mitigating the aggregation errors and zonal effect. For example, the mean within-tract commute distance for the most northeast tract in 2010 in Figure 3.1 was zero miles by the conventional centroid-to-centroid measure and it became 8.39 miles by the above simulation processes. The within-tract average commute distance became 7.74 miles after land use variability was accounted for.

---

* 75,000 is suggested for a balance between computation time and a sufficient sample size for OD trips.

CHAPTER **5**

# Commuting and Land Use

THIS CHAPTER MEASURES COMMUTING lengths in terms of both distance and time and examines the temporal trend of commuting lengths between 1990 and 2010. It then attempts to understand the observed commuting patterns. As we discussed in Sections 2.2 and 2.3, commuting may be explained by where they are (i.e., spatial factors) and who they are (i.e., nonspatial factors) (Wang, 2003). Being the most important spatial factor in existing studies, land use is tested in this chapter. Section 5.1 presents the principle to measure commuting lengths using the simulation approaches discussed in Sections 4.3 and 4.4. Section 5.2 analyzes the temporal trend of commuting patterns in Baton Rouge from 1990 to 2010. To understand the commuting patterns detected in Sections 5.1 and 5.2, we examine the relationship between it and three land use indices—distance from the central business district (CBD) in Section 5.3, the jobs–housing balance ratio in Section 5.4, and the job proximity index in Section 5.5. Interested readers may refer to Section 2.2 for a review of existing studies in the three land use indices. Similar to Chapter 4, each of the above topics is followed by case studies with step-by-step technical instructions.

## 5.1  MEASURING COMMUTING LENGTHS

In case studies throughout this book, commuting lengths are measured in both distance and time. Unlike commuting time, distance is not reported in the Census for Transportation Planning Packages (CTPP). Given the aforementioned issues, we retrieve commuting distance based on the simulation of individual trips for better accuracy.

People commute by multiple modes. It makes more sense to take mode choices into consideration when measuring commuting distance. For example, Wang (2000) recovered commute distance based on the centroid-to-centroid network distance in Chicago by two major modes—vehicles (including driving alone, carpooling, bus, and taxi) and trains (subway and rail) due to the high percentage in each mode. Specifically, he calculated the shortest-time network distance for commuting trips made by vehicles and train while using the Manhattan distance to estimate the commute distances for commuters using a bicycle or walking. Separation by different transportation modes is truly reasonable, especially for areas with high demand in each mode, and additional information such as the transit network would be needed. In Baton Rouge, however, the majority commuted by auto (including driving alone and carpooling), and the percentage was steady over time (i.e., 94–95 percent) (see Table 5.1). Therefore, we estimated commute distance based on the road network without taking transit, bicycle or pedestrian routes into consideration.

While commuting times are already provided in the CTPP, here we show the process to derive both commuting distances and times based on the simulated individual trips from Section 4.4 in Chapter 4 because of unreliability of reported travel time as discussed on page 17. Estimated commuting time is especially useful for a comparative measure of wasteful commuting in Chapter 6.

TABLE 5.1   Modal Splits and Commute Lengths in Baton Rouge 1990–2010

| | Modal Splits | | | | Mean Commute Distance (mile) | Mean Commute Time (min) |
|---|---|---|---|---|---|---|
| Year | Drove-Alone | Carpool | Public Transit | Others[a] | | |
| 1990 | 82.35 | 11.76 | 1.29 | 4.60 | 5.95 | 16.73 |
| 2000 | 83.16 | 11.91 | 1.40 | 3.54 | 6.17 | 18.73 |
| 2010 | 83.77 | 11.08 | 1.75 | 3.40 | 6.25 | 17.98 |

[a] Others include taxi, motorcycle, bicycle, walking, etc.

## Step 1   Preparing the Road Network Source Layer

The source feature dataset must have fields representing network impedance values such as travel distance and time through each road segment. ArcGIS recommends naming these fields as the units of impedance such as Meters (or Minutes) so that they can automatically be detected.

In ArcMap, open the attribute table of feature class rd _ 2010 stored in feature dataset rd _ 2010, and add fields Meters and Minutes (as type Float). Right-click the field Meters > Calculate Geometry > choose Length for Property and Meters for Units to calculate the length of each road segment. Given that the feature class rd _ 2010 already has a SPEED field specifying the moving speed on each road segment, right-click the field Minutes and calculate it as "=60*([Meters]/1609)/[SPEED]." Note that the unit of speed is miles per hour, and the formula returns the travel time in minutes. When no speed field is available, one can estimate the speed according to the U.S. Census Bureau's Census Feature Class Codes (CFCC) of roads. Specifically, add a field SPEED and assign a default value of 25 (mph) for all road segments, then update the values according to the following rules (Luo and Wang, 2003) by using a series of Select By Attributes and Field Calculator:

- For CFCC >= 'A11' and CFCC <= 'A18', Speed = 55

- For CFCC >= 'A21' and CFCC <= 'A28', Speed = 45

- For CFCC >= 'A31' and CFCC <= 'A38', Speed = 35

## Step 2  Building the Network Dataset in ArcGIS

In ArcCatalog, choose Customize from the main menu > Extensions > make sure that Network Analyst is checked, and close the Extensions dialog. This activates the Network Analyst module.

In ArcCatalog, navigate to your project folder, right-click the feature dataset ⊞ rd _ 2010 > New > Network Dataset. Follow the steps below to complete the process*:

- Name the New Network Dataset as rd _ 2010 _ ND and click Next.

- Select the feature class that will participate in the network dataset: rd _ 2010 and click Next.

- In the dialog window of modeling turns, choose No, and click Next.

- In the defining connectivity window, click Next.

- In the modeling elevation window, choose None as our dataset contains no elevation fields, and click Next.

- In the window for specifying the attributes, Network Analyst searches for and assigns the relevant fields: Meters, Minutes, and RoadClass > click Next.

- In the window for setting travel modes, click Next.

- In the window for establishing driving directions, choose No, and click Next.

- In the window for building service area index, leave it unchecked and click Next.

- Click Finish to close the summary window.

- Click Yes to build the new network dataset.

---

* The source feature dataset does not contain *reliable* network attributes such as turn or one-way restriction, elevation, accessible intersection, and so on. The above definitions are chosen for simplicity and also reduce the computation time in network analysis.

The new network dataset rd _ 2010 _ ND and a new feature class rd _ 2010 _ ND_Junctions become part of the feature dataset rd _ 2010. Right-click the network dataset rd _ 2010 _ ND > Properties, and review the Sources, Connectivity, Elevation, Attributes, and Directions defined above, and use Reset to modify any if needed.

### Step 3  Starting Network Analyst in ArcGIS

In ArcMap, to display the Network Analyst toolbar, choose Customize from the main menu > Toolbars > check Network Analyst.

Add the feature dataset rd _ 2010 to the Table Of Content (TOC) on the left of the main map window. All contents within it (rd _ 2010_ND _ Junctions, rd _ 2010, and rd _ 2010 _ ND) are displayed. For a faster display, one may only add the network dataset rd _ 2010 _ ND without other elements to the TOC. Also, add the text file simu _ trips.txt to the TOC, right-click it > Display XY Data > choose X for X Field and Y for Y Field, and OK to create a layer "simu _ trips.txt Events." Right-click on it and export it to a permanent feature class simu _ trips for subsequent use.

### Step 4  Computing the Shortest Paths between
### Origins and Destinations in ArcGIS

Active the New Route tool in the Network Analyst module. Next step is to define the Stops by loading the origins and destinations for all trips. Click 🖼 to open the Network Analyst window on the left of the TOC window. In the Network Analyst window, right-click the layer Stops (0) > Load Location > Select simu _trips as the Load From feature class; under Location Analysis Properties, specify the field for RouteName as Route _ ID; under Location Position, choose Use Geometry and set the Search Tolerance to 7,500 meters (one may adjust this setting to ensure the origins and destinations of every route are successfully located in the road network). Click OK to load points.

After loading the $2 \times 75{,}000 = 150{,}000$ locations (again, simu_trips save 150,000 records—the first half representing origins of 75,000 simulated trips and the second half denoting destinations of these trips), click the Solve button ▦ to compute the shortest paths for the 75,000 simulated O-D trips generated in Section 4.4. The result is saved in the layer Routes.

Though more accurate, the estimated trip lengths from such a great number of individual trips may not be the best way to capture and understand commuting patterns in a region. Instead, the mean commuting distance/time of a zone is commonly used in the literature as a measure of commuting patterns in the zone (Gera, 1979; Gera and Kuhn, 1980; Gordon et al., 1989a, 1991; Giuliano and Small, 1993; Cervero and Wu, 1998; Wang, 2000, 2001, 2003; Kim, 2008; Sultana and Weber, 2014). Following Wang (2000, 2001, 2003), we define the mean commuting distance/time in a tract as the average travel distance/time from this tract to all employment tracts weighted by the corresponding number of commuters.

$$MC_i = \sum_{j=1}^{n} \frac{f_{ij}}{R_i} c_{ij} \qquad (5.1)$$

In Equation 5.1, $MC_i$ is the mean commute distance/time in tract $i$; $f_{ij}$ is the commuter flow residing in tract $i$ and working in tract $j$; $c_{ij}$ is travel cost (i.e., distance/time) between tract $i$ and $j$; $R_i$ is the number of resident workers in tract $i$; and $n$ is the total number of tracts in the study area. In a word, this measure indicates that resident workers in tract $i$, on average, commute as far/long as $MC_i$. We measured the mean commute time in this manner by using the journey-to-work data from the CTPP part 3 including the reported commuter flow $f_{ij}$ and average commute time $c_{ij}$ (all travel modes). We have provided the corresponding two tables in the geodatabase br2010.gdb in the data folder. Table ctpp Flow stores the actual commuter flow $f_{ij}$, and table ctppTT saves the average all-travel-mode commute time $c_{ij}$ plus the number of resident workers for each tract $i$, $R_i$.

### Step 5 Joining `ctppFlow` to `ctppTT` Based on Two Common Fields in ArcGIS

Different from the traditional join process that works with only one field, this step involves two fields—the origin tract ID and destination tract ID. Therefore, the traditional join tool in ArcGIS is not applicable. Instead, go to ArcToolbox > Data Management Tools > Layers and Table Views > Make Query Table. Add `ctppTT` first and `ctppFlow` next to Input Tables. This particular order will identify `ctppTT` as the target table and `ctppFlow` as the source table. In the Fields list, one can choose what fields from both tables to keep since there are multiple duplicate fields. For example, check the fields `ctppTT.OBJECTID`, `ctppTT.O _ trtID _ new`, `ctppTT.O _ TotWrk`, `ctppTT.D _ trtID _ new`, `ctppTT.TT _ AllMeans`, and `ctppFlow.Flow` and leave the rest unchecked. In Expression, type `ctppFlow.O _ trtID _ new = ctppTT.O _ trtID _ new` AND `ctppFlow.D _ trtID _ new = ctppTT.D _ trtID _ new`. Click OK to get back to the Make Query Table dialog and click OK again to execute. The resulted table `QueryTable` is automatically added to the TOC. Export it as `ctppTT _ Flow` for subsequent use.

### Step 6 Calculating the Mean Commuting Time for a Tract in ArcGIS

Open `ctppTT _ Flow` and add a new field `Comm _ Time` as type Float. Use Field Calculator to update it by setting `Comm _ Time = ctppFlow _ Flow * ctppTT _ TT _ AllMeans / ctppTT _ O _ TotWrk`. Next, right-click on `O _ trtID _ new` and select Summarize. In step 2 in the Summarize window, unfold `Comm _ Time` and check Sum. Specify the name of the output table as `Mean _ Comm _ Time` in step 3 and click OK. The field `Sum _ Comm _ Time` in `Mean _ Comm _ Time` stores the mean commuting time (in minutes) for each census tract.

Mean commute distance, while not reported, was measured based on the above Monte Carlo simulation of individual trips in Section 4.4. Equation 5.1 is now revised as Equation 5.2 for simulated trips. Specifically, if a simulated trip $k$ starts in tract $i$, $f_k$ then equals 1, indicating one eligible trip; $c_k$ is the network distance measured for trip $k$; $m$ denotes the total number of simulated trips.

$$MC_i = \sum_{k=1}^{m} \left( \frac{f_k c_k}{R_i} \right) \tag{5.2}$$

where $f_k = 1$ if trip $k$ starts in tract $i$, 0 otherwise.

### Step 7 Determining the Origin Tract and Destination Tract for Each Simulated Trip in ArcGIS

Add `trt` to the TOC. Select the first 75,000 records in the attribute table of `simu _ trips` and then right-click on `simu _ trips` in the TOC to export the selected features as a new feature class `O _ indiv`. Then go back to the attribute table of `simu _ trips` and click Switch Selection to select the last 75,000 records, that is, all the destinations of the simulated trips. Repeat the previous step and save them into a new feature class `D _ indiv`. Remove `simu _ trips` to wipe it from the memory. Then right-click on `O _ indiv` > Joins and Relates > Join. In the Join Data window, choose "Join data from another layer based on spatial location" in What do you want to join to this layer? Choose `trt` in step 1, select "it falls inside" in step 2, and save the result as a new feature class `O _ indiv _ trt`. Click OK to execute. This will attach `trtID _ new` to each of the 75,000 trip origins. Right-click on `trtID _ new` to change its alias to `O _ trtID _ new`. Likewise, repeat the process for `D _ indiv` to find out destination tracts for each trip and change the alias of `trtID _ new` to `D _ trtID _ new`. The steps to alter the alias of fields are to avoid confusion in the next step.

### Step 8 Joining the Origin Tract ID and Destination Tract ID to the Trip Length for Each Trip

Add `Routes` generated from step 4 in this section to the TOC. Right-click on `Routes` to join `O _ indiv _ trt` to it.

As O _ indiv _ trt has several irrelevant fields, one may delete those fields before the join. Next, join D _ indiv _ trt to Routes. Then export the attribute table of Routes into a table Routes _ trt. Remove irrelevant fields in Routes _ trt so that it has only Name (i.e., Route _ ID), Total _ Length, O _ trtID _ new, and D _ trtID _ new.

### Step 9 Calculating the Mean Commute Distance for a Tract

Open the attribute table of Routes _ trt. Right-click on the field O _ trtID _ new and select Summarize. In step 2 in the Summarize window, unfold Total _ Length and check Sum. Specify the name of the output table as Mean _ Comm _ Dist in step 3 and click OK. The field Sum _ Total _ Length in Mean _ Comm _ Dist stores the mean commuting distance (in meters) for each census tract.

## 5.2 OVERALL TEMPORAL TREND OF COMMUTING

Following the above procedures, we also calculated the mean commuting distances and times for 1990 and 2000 and reported the results in Table 5.1. We find that the mean commute distance on average increased steadily from 5.95 miles in 1990 to 6.17 miles in 2000 and further to 6.25 miles in 2010. This is consistent with some previous studies (Levinson and Kumar, 1994; Cervero and Wu, 1998) and reflects a long-standing trend of more workers moving farther from their jobs either to search farther for jobs for maximizing their earnings or to move their residences for better housing. However, the increasing rate was higher in 1990–2000 than in 2000–2010. One possible reason might be the recession in 2008 (Horner and Schleith, 2012).

The mean commute time for the overall population increased from 1990 to 2000 and then had declined by 2010. Other studies also found that the average commute time stayed stable or even dropped over time, albeit the worsening traffic congestion (Gordon et al., 1989c, 1991; Dubin, 1991; Levinson and Kumar, 1994; Kim, 2008). They ascribed the declining time to the colocation of jobs and housing, that is, people relocate their residence or jobs to cut back commute time as traffic becomes more congested (refer to Section 2.3 in Chapter 2 for more detail). An increasing use of suburban roads that are usually newer and wider than roads in the central city makes it achievable to commute a longer distance in a shorter time. Taking driving alone as an example, the implied average commuting speed was 24.3 mph in 1990, which dropped to 21.9 mph in 2000, and only recovered slightly to 22.8 mph in 2010. Therefore, a small increment in commute distance from 1990 to 2000 came with a much larger climb in commute time, and it was only after 2000 that the colocation theory became relevant and led to a small drop in commute time. The significant climb of commute time for 1990–2000 might reflect people's increasing endurance of long commutes or traffic congestion that grew much more rapidly than people could adapt (Levinson and Wu, 2005), and the relocation adjustment came afterward. Such an increase of commuting time in 1990–2000 was also seen in Pisarski (2002).

Based upon the observed commuting patterns (both distance and time) as listed in Table 5.1, we may ask questions of why and how. In short, we need to find the underlying factors that significantly affected commuting. This chapter focuses on the spatial part by looking at the land use layout, that is, the spatial distributions of employment and resident workers. As mentioned in Section 2.2 in Chapter 2, existing attempts look at three aspects of land use patterns—the distance from the CBD, the local jobs–housing balance ratio, and the overall proximity to jobs—and link them with commuting. Specifically, the distance from the CBD captures how far a residential location is from a job concentration area (particularly applied to a monocentric city where its CBD dominates the job market); the local jobs–housing balance ratio further extends the measure by considering local jobs (within a threshold of residence) rather than jobs in one particular area (e.g., CBD); and the job proximity, collectively, measures the distance from a residential location to the overall job market. The effects of the three land use metrics on commuting will be examined separately in the following sections.

## 5.3 COMMUTING PATTERN VS. DISTANCE FROM THE CBD

There has been a long tradition of attempts to explain intra-urban variability of commuting by land use patterns. The analysis begins by examining the impact of the CBD with the highest employment concentration (see Figure 3.1). In Baton Rouge, we found that 42 percent of the workers commuted to the area within a 3-mile radius of the CBD in 1990, and the rate dropped to 32 percent in 2000, and 31 percent in 2010, indicating that a large portion of jobs were concentrated in the CBD area albeit they increasingly exited the CBD during this period. The simple scatter plot in Figure 5.1a shows an obvious positive relationship (nearly linear) between a tract's distance from the CBD and either its mean commute distance or time. A rigorous statistical analysis is then followed to test the significance of the observed positive relationship.

### Step 10 Calculating a Tract's Distance from the CBD in ArcGIS

Add the point feature class BRCenter, which represents the CBD of Baton Rouge, and another point feature class trtpt to the TOC. In ArcToolbox, choose Analysis Tools > Proximity > Point Distance. Select trtpt as Input Features and BRCenter as Near Features. Name the Output Table as Dist _ from _ CBD and click OK to execute. The field DISTANCE in the resulted table describes the Euclidean distances from each tract center to the CBD.

### Step 11 Running a Regression Model on a Tract's Distance from the CBD and Its Mean Commute Distance and Time

Add the table Mean _ Comm _ Dist from step 9 in Section 5.1 to the TOC. Then right-click on it and select Joins and Relates > Join to join Dist _ from _ CBD to this table. In the Join Data window, choose O _ trtID _ new in step 1, Dist _ from _ CBD in step 2, and INPUT _ FID in step 3. Click OK to perform the join.

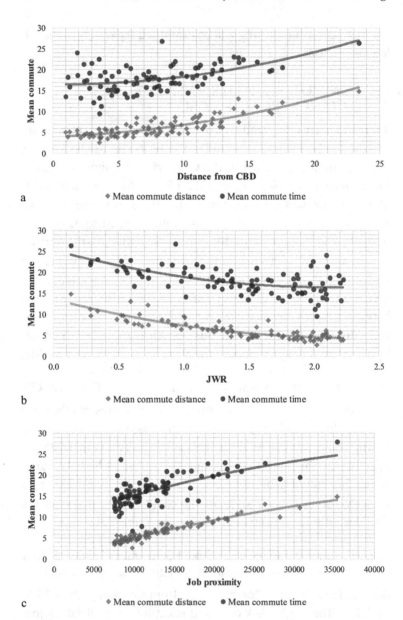

FIGURE 5.1   Mean commute distance and time vs. land use metrics in 2010: (a) distance from the CBD, (b) jobs–housing balance, and (c) job proximity.

From the Table Options menu click Export and save the joined table as DCBD _ CommDist.dbf (as type dBASE Table). Use Excel to open the file. Note that the names of some fields may be truncated as a result of the ten-character limit of a field name in a dbf file. For example, the mean commute distance field Sum _ Total _ Length becomes Sum _ Total _ in our case. In Excel, go to the Data Analysis tool in the Analysis group on the Data tab. If not present, one needs to load the Analysis ToolPak add-in to Excel by (1) clicking the File tab > Options > Add-Ins, (2) selecting Excel Add-ins in the Manage box and clicking Go, and (3) checking the Analysis ToolPak check box in the Add-ins box and clicking OK. In the Data Analysis window, select Regression and click OK. Select the records in Sum _ Total _ column as Input Y Range and the records in DISTANCE as Input X Range. Click OK to run the regression model. Results including the coefficient of the independent variable, its t-statistics and p-value, and the model explanation power $R^2$ are reported in Table 5.2. It should be noted that one may use other statistical software such as R, SAS, and SPSS to perform the regression analysis. Repeat the above process for the mean commute time. Results are listed in Table 5.2 as well.

Table 5.2 shows that the distance from the CBD (denoted by $D_{CBD}$) explained the variation of mean commute distance across census tracts by 78 percent in 1990, 68 percent in 2000, and 63 percent in 2010. The declining explaining power was attributable to the dispersion of jobs beyond the CBD area (e.g., percentage of workers commuting to the area within a 3-mile radius of the CBD was 42 percent in 1990, 32 percent in 2000, and 31 percent in 2010). The effect of the distance from the CBD remained significant on mean commute time in all models (Table 5.3), but much weaker than on mean commute distance (Table 5.2). The lower performance of regression models on commute time was attributable to the non-uniform modal distribution across tracts. In fact, for commuters driving alone, the pattern of mean commute time was largely consistent with that of mean commute distance, and the corresponding regression models yielded $R^2 = 0.60$, 0.41, and

TABLE 5.2  Regression Models of Mean Commute Distance across Census Tracts 1990–2010

| | 1990 | 2000 | 2010 | 1990 | 2000 | 2010 | 1990 | 2000 | 2010 |
|---|---|---|---|---|---|---|---|---|---|
| Intercept | 2.40*** (10.03) | 2.91*** (10.52) | 3.01*** (9.93) | 10.37*** (29.14) | 14.87*** (20.75) | 13.65*** (21.34) | 0.88*** (4.60) | 0.99*** (7.06) | 1.10*** (5.74) |
| $D_{CBD}$ | 0.57*** (17.38) | 0.50*** (13.67) | 0.48*** (12.22) | | | | | | |
| JWR | | | | −3.88*** (−14.07) | −10.38*** (−8.35) | −8.05*** (−7.57) | | | |
| JWR$^2$ | | | | 0.37*** (13.93) | 2.60*** (5.33) | 1.75*** (4.36) | | | |
| JobP | | | | | | | 0.67*** (29.44) | 0.67*** (40.26) | 0.64*** (29.22) |
| No. obs. | 85 | 89 | 91 | 85 | 89 | 91 | 85 | 89 | 91 |
| $R^2$ | 0.78 | 0.68 | 0.63 | 0.72 | 0.78 | 0.76 | 0.91 | 0.95 | 0.91 |

*Note:* t-statistics are in parentheses; *** significant at the 0.001 level.

TABLE 5.3 Regression Models of Mean Commute Time across Census Tracts 1990–2010

| | 1990 | 2000 | 2010 | 1990 | 2000 | 2010 | 1990 | 2000 | 2010 |
|---|---|---|---|---|---|---|---|---|---|
| Intercept | 14.72*** (26.82) | 16.96*** (23.06) | 15.29*** (25.01) | 19.69*** (28.25) | 27.86*** (13.51) | 25.36*** (17.12) | 13.32*** (20.44) | 14.42*** (16.12) | 13.63*** (21.15) |
| $D_{CBD}$ | 0.32*** (4.28) | 0.27** (2.80) | 0.40** (5.04) | | | | | | |
| JWR | | | | -2.51*** (-4.65) | -12.26*** (-3.43) | -8.45*** (-3.43) | | | |
| JWR² | | | | 0.21*** (4.12) | 3.60* (2.57) | 1.99* (2.15) | | | |
| JobP | | | | | | | 0.25*** (5.74) | 0.30*** (5.22) | 0.31*** (7.41) |
| No. obs. | 85 | 89 | 91 | 85 | 89 | 91 | 85 | 89 | 91 |
| R² | 0.18 | 0.08 | 0.22 | 0.21 | 0.26 | 0.35 | 0.28 | 0.24 | 0.38 |

*Note:* t-statistics are in parentheses; * significant at the 0.05 level, ** significant at the 0.01 level, *** significant at the 0.001 level.

0.50 in 1990, 2000, and 2010, respectively (details not reported here). Adding the square term $D_{CBD}^2$ did not improve the explaining power of the regression models and thus was not included. Note that results on jobs-to-workers ratio (JWR) and proximity to jobs (JobP) are also reported in Tables 5.2 and 5.3; see the following sections for more details.

## 5.4 COMMUTING PATTERN VS. JOBS– HOUSING BALANCE

The metric distance from the CBD only considers jobs in the CBD area. This section further considers jobs within a certain range of residential locations, or the so-called jobs–housing balance ratio (Cervero, 1989). An imbalanced area has far more resident workers than jobs, and thus more workers need to commute outside of the area for their jobs and tend to incur more commuting. The "area" can be defined by different shapes such as squares (Peng, 1997) and circles (Wang, 2000). Following Wang (2000), this research used the floating catchment area method to define a circular area around each census tract centroid and calculated the jobs-to-workers ratio (JWR) within each catchment area. A higher JWR implies less need for commuting beyond the catchment area and thus is expected to correlate with less commute. We experimented with radii ranging 1.5–7.5 miles and settled with 5 miles for its best explaining power.

Step 12   Generating 5-Mile Catchment
                Areas for Tracts in ArcGIS

Add `trtpt` to the TOC. Go to ArcToolbox > Analysis Tools > Proximity > Buffer. Set `trtpt` as Input Features. Set Output Feature Class as `trtpt _ catchment`. In the Linear unit textbox under Distance, put 8,046.72 meters (the unit of the projected coordinate system is meter). Click OK.

Step 13   Calculating JWR for Tracts

Following step 7 in Section 5.1, use spatial join to find the number of available jobs within the 5-mile catchment area of each residential tract. Specifically, in the Join Data window—again, make sure "Join data from another layer based on spatial location" is selected—choose trtpt in step 1, check Sum under the first option in step 2, "Each polygon will be given a summary of the numeric attributes of the points that fall inside it, and a count field showing how many points fall inside it." Name the output feature class as JWR _ 5mile. Open the attribute table of it, add a new field (as type Float), and calculate the field by dividing Sum _ TotJob by Sum _ TotWrk.

Following step 11 in Section 5.3, we ran regression models on the relationships between a tract's JWR and its mean commuting length (distance and time) for all three time periods. As an example, Figure 5.1b shows the relationship between mean commute distance and time versus JWR in 2010. Both display a quadratic trend, but the trend is much clearer for distance than time. This is confirmed by a regression model with the added square term of JWR, as reported in Tables 5.2 and 5.3. In 1990, 2000, and 2010, the negative sign of JWR and the positive sign of $JWR^2$ indicate that mean commute distance or time at the tract level declined with JWR, but the declining slope got flatter in higher-JWR areas. Both terms are statistically significant in all models. The models for mean commute distance in Table 5.2 performed well with $R^2$ that was 0.72 in 1990, peaked at 0.78 in 2000, and dropped slightly to 0.76 in 2010. The models for mean commute time in Table 5.3 also confirmed the quadratic trend in all years though with lower $R^2$ (0.21, 0.26, and 0.35 in 1990, 2000, and 2010, respectively). In short, the results confirm the importance of the jobs–housing imbalance affecting commuting patterns, which are much more significant in Baton Rouge than large metropolitan areas reported in other studies (e.g., Wang, 2000).

## 5.5 COMMUTING PATTERN VS. PROXIMITY TO JOBS

Either the emphasis on the role of the CBD or the jobs–housing balance approach does not consider all job locations in explaining the commuting patterns. The job proximity index (JobP) captures the spatial separation between a worker's residence location and all potential job sites (Wang, 2003), formulated such as:

$$JobP_i = \sum_{j=1}^{n} P_{ij} d_{ij} \qquad (5.3)$$

where
$$P_{ij} = \left( J_j d_{ij}^{-\beta} \right) \bigg/ \sum_{k=1}^{n} \left( J_k d_{ik}^{-\beta} \right)$$

Similar to the notion of Huff's (1963) model, the probability of workers residing in zone $i$ and going to work in zone $j$ (denoted by $P_{ij}$) is predicted as the gravity kernel of job site $j$ out of those of all job sites $k$ (=1, 2, …, $n$). Each gravity kernel is positively related to the number of jobs there $J_j$ (or $J_k$) and negatively to the distance or time (measured as network distance or time) between them $d_{ij}$ (or $d_{ik}$) powered to the distance friction coefficient $\beta$. With the probability $P_{ij}$ defined, JobP at zone $i$ is simply the aggregation of all distances (time) $d_{ij}$ with corresponding probabilities $P_{ij}$ over all job sites ($j = 1, 2, …, n$).

Calibration of JobP requires defining the value for the distance friction coefficient $\beta$ in Equation 5.3. Here, the $\beta$ value was computed from the log-transformed regression based on the classic gravity model such as

$$C_{ij} = aW_i J_j d_{ij}^{-\beta} \qquad (5.4)$$

where $C_{ij}$ is the number of commuters from a tract with $W_i$ resident workers and to a tract with $J_j$ jobs for a distance (time) of $d_{ij}$ (Wang, 2015). Based on the CTPP data, the derived $\beta$ value was 0.404 in 1990, 0.547 in 2000, and 0.475 in 2010 if the journey-to-work trips were measured in distance; and the $\beta$ value was 0.295 for 1990, 0.353 for 2000, and 0.385 for 2010 if measured in time. Here, we only show the process to calculate the $\beta$ value in terms of distance and JobP in 2010 as an example.

## Step 14 Deriving Tract-to-Tract Commuting Trips Based on Simulated Individual Trips

Recall that the attribute table of Routes _ trt from step 8 in Section 5.1 records the origin tract, destination tract, and travel distance for each simulated individual trip. One possible way to aggregate individual level trips to tract level ones is to use Summarize. However, the Summarize tool available in the attribute table only deals with one single field. To calculate summary statistics based on two fields—O _ trtID _ new and D _ trtID _ new—we need another tool. It can be accessed in ArcToolbox > Analysis Tools > Statistics > Summary Statistics. Select Routes _ trt as the Input Table. Name Output Table as Trt _ Dist. From the Statistics Field(s) dropdown list, select Total _ Length and select MEAN as the Statistic Type. This will calculate the average commute distances between every two tracts (it is not the same as the mean commute distance for a tract in step 9 in Section 5.1). In the Case field, select O _ trtID _ new and D _ trtID _ new. Click OK. The resulted table Trt _ Dist has four fields—O _ trtID _ new, D _ trtID _ new, FREQUENCY, and MEAN _ Total _ Length, representing the origin tract ID, destination tract ID, number of trips between them, and average commute distances between them, respectively. It should be noted that the FREQUENCY field does not report the actual number of commuters between two tracts, since the calculation is based on 75,000 simulated trips. One may doublecheck this by right-clicking on FREQUENCY and clicking Statistics to observe the sum statistics. Next, we will follow step 5 in Section 5.1 to join Trt _ Dist to ctppFlow (the field Flow in ctppFlow records the actual commuting flow between tracts). Some entries in the field MEAN _ Total _ Length could be null since all possible tract pairs ($91*91=8{,}281$) are included in ctppFlow while only tract pairs with actual commuting trips in Trt _ Dist. Select those records and change their values from null to 0. Save the table as ctppFlow _ Dist.

Step 15  Joining the Number of Workers of Origin
Tracts ($W_i$) and Number of Jobs of Destination
Tracts ($J_j$) to ctppFlow _ Dist

Add trtpt to the TOC. Join it to ctppFlow _ Dist based on
the common field, tract ID (trtID _ new for trtpt and O _
trtID _ new for ctppFlow _ Dist). Save it as a permanent
table and change the name of field TotWrk to O _ TotWrk. Then
join trtpt to the newly created table based on D _ trtID _
new and export it as Dist _ decay. Change the name of the
second TotJob field to D _ TotJob (the first TotJob field
belongs to the origin tract). The Dist _ decay table has Flow,
O _TotWrk, D _TotJob, and MEAN _ Total _ Length,
which corresponds to $C_{ij}$, $W_i$, $J_j$, and $d_{ij}$ in Equation 5.4, respectively. Export Dist _ decay to a dbf file.

Step 16  Calculating the Distance Decay
Friction Coefficient β in Excel

Open the Dist _ decay.dbf file using Excel and compute a
new field C/WJ by dividing Flow by the product of O _ TotWrk
and D _ TotJob. Add an X–Y scatter graph depicting how $d_{ij}$
(MEAN _ Total _ Length) varies with $C_{ij}/W_jJ_j$ (C/WJ). Click
the data points on the graph > Add Trendline and choose the
Power trendline option. Check the options "Display Equation on
chart" to have the full equations including the estimated β value
added to the graph.

Step 17  Calculating the Job Proximity Index for a Tract

Open Dist _ decay (not the dbf file) in ArcMap and add a
new field Jd (as type Float) = D _ TotJob * (MEAN _ Total _
Length ^ −0.475). Recall that −0.475 is the distance decay friction coefficient in terms of distance in 2010. This calculates the
numerator of $P_{ij}$ in Equation 5.3. Summarize on O _ trtID _
new and calculate the Sum of Jd. This will create a field Sum _
Jd, which calculates the denominator of $P_{ij}$ in the new table after
Summarize. Then join the new table to Dist _ decay based on

the common field O _ trtID _ new. Add a new field Pij (as type Float) = Jd/Sum _ Jd. Again, add a new field Pijdij (as type Float) = Pij * MEAN _ Total _ Length. Finally, summarize on O _ trtID _ new and calculate the Sum of Pijdij. Name and save the new table as JobP _ Dist. In the new table, the field Sum _ Pijdij describes the JobP in terms of distance for each tract. Repeat the entire process and calculate the JobP in terms of travel time for each tract.

Follow step 11 in Section 5.3 to join JobP _ Dist to Mean _ Comm _ Dist, draw an X–Y scatter plot (see Figure 5.1c), and run a regression analysis. The regression results for mean commute distance and time by JobP are again reported in Tables 5.2 and 5.3, respectively. The mean commute distance at the tract level was well explained by JobP with $R^2$ = 0.91, 0.95, and 0.91 in 1990, 2000, and 2010, respectively. This is a significant improvement over the other two factors $D_{CBD}$ and JWR. Similarly, regression models on mean commute time returned lower $R^2$ ranging from 0.24 to 0.38. We also did not add the square term $JobP^2$ as it did not improve the explaining power of the regression models. Note that we also ran a series of regression models on mean commuting distance (centroid-to-centroid), and results indicate weaker $R^2$ than the simulation-calibrated distance. This again demonstrates the value of our simulation approaches.

# CHAPTER 6

# Wasteful Commuting

LTHOUGH THE LAND USE pattern in Baton Rouge explained the commuting length variations to some extent, there was still a proportion of variations that were unexplained (particularly commuting time). Part of the gap could be attributable to the wasteful commuting issue that individuals do not necessarily optimize their journey-to-work trips as suggested by the spatial arrangements of land uses, that is, jobs and houses (Hamilton, 1982). Even if individual workers make every effort to minimize their commuting, the outcome may still differ from the minimum total commute collectively for the whole study area.

Wasteful commuting $T_w$ is the proportion of the average actual commute $T_a$ that is over the average required commute $T_r$, that is,

$$T_w = (T_a - T_r)/T_a \qquad (6.1)$$

Figure 6.1 outlines the workflow of our analysis in this chapter. Based on the 2006–2010 Census for Transportation Planning Packages (CTPP), we first replicate the existing approach of measuring wasteful commuting at the zonal (census tract) level, then we use the Monte Carlo method to simulate individual locations of resident workers and employment as well as trips between

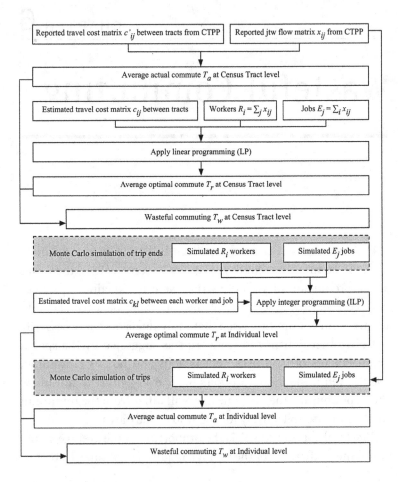

FIGURE 6.1 Workflow of measuring wasteful commuting at the zonal and individual levels.

them, and finally we calibrate wasteful commuting at the individual level. Section 6.1 introduces the classic mathematical model in the seminal work by Hamilton (1982). Section 6.2 then describes an improved and simplified optimization model proposed by White (1988). Section 6.3 presents a case study with detailed technical instructions on how to measure wasteful commuting at the census tract level. For comparison, Section 6.4

showcases another case study of measuring wasteful commuting at the individual level, the most disaggregate level that a study could achieve.

## 6.1 HAMILTON'S MODEL ON WASTEFUL COMMUTING

Economists often make assumptions in order to simplify a model with reasonable complexity while capturing the most important essence of real-world issues. Similar to the monocentric urban economic model, Hamilton (1982) made some assumptions for the urban structure. One also needs to note two limitations to Hamilton in the early 1980s: the lack of intra-urban employment distribution data at a fine geographic resolution and the GIS technology at its developmental stage.

First, consider the commuting pattern in a monocentric city where all employment is concentrated at the central business district (CBD) (or the city center). Assume that the population is distributed according to a density function $P(x)$, where $x$ is the distance from the city center. The concentric ring at distance $x$ has an area size $2\pi x dx$, and thus population $2\pi x P(x)dx$, who travels a distance $x$ to the CBD. Therefore, the total distance $D$ traveled by a total population $N$ in the city is the aggregation over the whole urban circle with a radius $R$:

$$D = \int_0^R x(2\pi x P(x))dx = 2\pi \int_0^R x^2 P(x)dx \qquad (6.2)$$

Therefore, the average commute distance per person $A$ is

$$A = \frac{D}{N} = \frac{2\pi}{N} \int_0^R x^2 P(x)dx \qquad (6.3)$$

Now assume that the employment distribution is decentralized across the whole city according to a function $E(x)$. Hamilton believed that this decentralized employment pattern was more realistic than the monocentric one. Similar to Equation 6.3, the average distance of employment from the CBD is

$$B = \frac{2\pi}{J} \int_0^R x^2 E(x) dx \qquad (6.4)$$

where $J$ is the total number of employment in the city.

Assuming that residents can freely swap houses in order to minimize commutes, the planning problem here is to minimize total commuting given the locations of houses and jobs. Note that employment is usually more centralized than population. The solution to the problem is that commuters always travel toward the CBD and stop at the nearest employer. Comparing to a monocentric city, "displacement of a job from the CBD can save the worker a commute equal to the distance between the job and the CBD" (Hamilton, 1982, p. 1040). Therefore, "optimal" commute or *required commute* or *minimum commute* per person is the difference between the average distance of population from the CBD ($A$) and the average distance of employment from the CBD ($B$):

$$C = A - B = \frac{2\pi P_0}{N} \int_0^R x^2 e^{-tx} \, dx - \frac{2\pi E_0}{J} \int_0^R x^2 e^{-rx} dx \qquad (6.5)$$

where both the population and employment density functions are assumed to be exponential, that is, $P(x) = P_0 e^{-tx}$ (see Section 4.2), and $E(x) = E_0 e^{-rx}$, respectively.

Solving Equation 6.5 yields

$$C = -\frac{2\pi P_0}{tN} R^{-2} e^{-tR} + \frac{2}{t} + \frac{2\pi E_0}{rJ} R^{-2} e^{-rR} - \frac{2}{r} \qquad (6.6)$$

Hamilton studied 14 American cities of various sizes and found that the required commute only accounts for 13 percent of the actual commute, and the remaining 87 percent is wasteful. He further calibrated a model in which households choose their homes and job sites at random and found that the random commute distances were only 25 percent over the actual commuting distances, much closer than the optimal commute!

There are many reasons why people commute more than the required commute predicted by Hamilton's model. Some are recognized by Hamilton himself, such as bias in the model's estimations (residential and employment density functions, radial road network)* and the assumptions made. For example, residents do not necessarily move close to their workplaces when they change jobs because of relocation costs and other concerns. Relocation costs are likely to be higher for homeowners than renters, and thus homeownership may affect commute. There are also families with more than one worker. Unless the jobs of all workers in the family are at the same location, it is impossible to optimize commute trips for each income earner. More importantly, residents choose their homes for factors not related to their job sites such as accessibility to other activities (shopping, services, recreation and entertainment, etc.), quality of schools and public services, neighborhood safety, and others. Some of these factors are considered in research in explaining the intra-urban variation of actual commuting (e.g., Shen, 2000; Wang, 2001).

## 6.2 WHITE'S MODEL ON WASTEFUL COMMUTING

Hamilton (1982) used exponential functions to approximate the spatial distribution of homes and jobs and the straight-line distance from the CBD to estimate the commuting lengths. As mentioned in Section 2.4, White (1988) argued that the urban commute optimization should be constrained to the existing

---

* Hamilton did not differentiate residents (population in general including dependents) and resident workers (those actually in the labor force who commute). By doing so, it was assumed that the labor participation ratio was 100 percent and uniform across an urban area.

spatial distribution of homes and jobs and the road network, and she formulated the optimal commuting pattern by a simple Linear Programming (LP) approach. Its formulation is shown as follows:

$$\text{Minimize} \quad T_r = \sum_i \sum_j \left( c_{ij} x_{ij} \right) \big/ N \tag{6.7}$$

$$\text{Subject to:} \quad \sum_j x_{ij} = R_i \tag{6.8}$$

$$\sum_i x_{ij} = E_j \tag{6.9}$$

$$x_{ij} \geq 0 \tag{6.10}$$

where $c_{ij}$ is the commuting time from zone $i$ to zone $j$; $x_{ij}$ is the number of commuters living in zone $i$ and working in zone $j$; $R_i$ represents the number of commuters living in zone $i$; $E_j$ represents the number of commuters working in zone $j$; $N = \Sigma_i \Sigma_j x_{ij}$ is the total number of commuters.

Equation 6.7 defines the objective function of minimizing the average commuting time, which is subject to three constraints. Equation 6.8 ensures that all journey-to-work flows originating from one zone satisfy the total number of commuters living in that zone. Similarly, Equation 6.9 limits all journey-to-work flows ending in one zone to the total number of commuters working in that zone. Equation 6.10 restricts the journey-to-work flow matrix $x_{ij}$ to be non-negative values. The resulting $T_r$ returned by the objective function represents the average required commuting time, suggested by the spatial arrangement of homes and jobs. Comparing required commute $T_r$ to actual commute $T_a$, Equation 6.1 measures the wasteful commuting rate.

## 6.3 MEASURING WASTEFUL COMMUTING AT THE CENSUS TRACT LEVEL

As stated previously, both travel time and distance are used to measure wasteful commuting. For a demonstration of the methods, we only used travel time. We adopted the popular LP technique to derive the optimal commute. Commonly referred to as the "transportation problem," the goal is to solve the optimal journey-to-work flows between origin and destination zones in order to minimize the average travel time (Hitchcock, 1941; Horner and Murray, 2002; Taaffe et al., 1996), such as presented in Equations 6.7 through 6.10. Similar to other case studies in the book, the following analyses use the 2010 data for Baton Rouge for demonstration.

A key parameter to the calculation of wasteful commuting is intrazonal (within-zone) travel distance or time. In implementation, intrazonal travel distance $c_{ii}$ is approximated as the radius of the minimum bounding circle of a zone (Frost et al., 1998; Horner and Murray, 2002), and the corresponding intrazonal travel time is obtained by simply assuming a constant travel speed of 25 mph through the distance. The interzonal travel distance (time) $c_{ij}$ is calibrated between the centroids of census tracts by the shortest network path, and then modified by adding the intrazonal components (distance or time) at both the origin and destination zones.

### Step 1  Generating the Minimum Bounding Circle of a Tract

Add the polygon feature class trt to the TOC. Go to ArcToolbox > Data Management Tools > Features > Minimum Bounding Geometry. Set trt as Input Features. Name the output feature class as trt _ MBC. Choose CIRCLE in the Geometry Type. Check "Add geometry characteristics as attributes to output" to have the diameter of the circle (saved in the field MBG _ Diameter) in the output feature class. Click OK to run.

Step 2  Measuring Interzonal Travel Times
(and Distances) between Tracts

Step 4 in Section 5.1 in Chapter 5 discussed the procedures for identifying the shortest path and calculating its distance or time for each given route using the New Route tool. Here, we will need another network analysis tool—OD cost matrix—to measure the travel times (or distances) between all possible OD pairs. A critical step to generate the OD time matrix is to prepare a network dataset with proper impedance (time) of each road segment. This was covered in steps 1–2 in Section 5.1 in Chapter 5.

*Step 2.1  Activating the OD Cost Matrix Tool*
On the Network Analyst toolbar, click the Network Analyst drop-down menu and choose New OD Cost Matrix. A composite network analysis layer "OD Cost Matrix" (with six empty classes: Origins, Destinations, Lines, Point Barriers, Line Barriers, and Polygon Barriers) is added to the layers window under the TOC. The same six empty classes are also added to the Network Analyst window under "OD Cost Matrix" (if not shown, click the icon 🖼 next to the Network Analyst dropdown menu to activate it).

*Step 2.2  Defining Origins and Destinations*
In the Network Analyst window under "OD Cost Matrix,"

- Right-click Origins (0) > Load Locations > In the dialog window, for Load From, choose `trtpt`; for both Sort Field and Name, select `trtID _ new`; for Location Position, choose Use Geometry and set the Search Tolerance 5,000 meters*; and click OK—91 origins are loaded.

---

* In essence, a route between an origin and a destination is composed of three segments: the segment from the origin to its nearest junction on the network (as the crow flies), the segment from the destination to its nearest junction on the network (as the crow flies), and the segment between these two junctions through the network. The default search tolerance (5,000 m) may not be applicable to other study areas.

- Right-click Destination (0) > Load Locations > In the dialog window, similarly, choose `trtpt` for Load From, select `trtID _ new` for both Sort Field and Name, and set the same search tolerance 5,000 under Use Geometry, and click OK—91 destinations are loaded.*

*Step 2.3 Computing the OD Cost (Travel Time) Matrix*

On the Network Analyst toolbar, click the Solve button. The solution is saved in the layer Lines under the TOC or in the Network Analyst window. Right-click either one > Open Attribute Table. The table contains the fields `OriginID`, `DestinationID`, and `Total _ Minutes`, representing the origin's ID (consistent with field `trtID _ new` in the feature class `trtpt`), the destination ID (consistent with field `trtID _ new` in the feature class `trtpt`), and total minutes between them, respectively.

On the above open table, click the Table Options dropdown icon > Export and name the table `ODTime`. Similarly, obtain `ODDist` for interzonal distances.

## Step 3 Adjusting the Travel Times (and Distances) between Tracts

The total travel time between two tracts is composed of the aforementioned network time and the intrazonal time at both the origin and destination tracts. The intrazonal time is approximated as the radius of a tract's minimum bounding circle divided by a constant speed 670.56 meters/minute (i.e., 25 mph) in this case study. Add a new field `intra _ time` (as type Float) in the attribute table of feature class `trt _ MBC` and calculate it as `0.5 * MBG _ Diameter/670.56`. For intrazonal travel distance, simply add another field `intra _ dist` and calculate it as `0.5 * MBG _ Diameter`. Then join the attribute table of feature class `trt _ MBC` to the attribute table of `ODTime` based on the common fields

---

* Do not load the origins and destinations multiple times, as it will add duplicated records to the analysis.

trtID _ new and OriginID, and then join the attribute table of feature class trt _ MBC to the above-joined table based on the common fields trtID _ new and DestinationID. By doing so, the intrazonal travel times for both the origin and destination tracts are attached to the interzonal travel time table. Add a new field NetwTime to the newly joined table and calculate it by summing the two intrazonal times and one interzonal time together. This amends the network travel time by adding intrazonal times at both trip ends. Export it to a text file odtime.txt to preserve the expanded table. Simplify the text file odtime.txt by keeping only three columns of values with headings such as Oid (ids for the origins or resident workers), Did (ids for the destinations or employment), and Time (adjusted travel time in minutes). This file is provided in the data folder for convenience. Repeat the process to obtain the travel distance table oddist.txt, which is also provided along with the book.

The fields Res and Emp in wrk _ job _ inBR _ 2010.txt (provided in the data folder) define the number of employment and resident workers in each tract,* respectively, and thus the variables $R_i$ ($i = 1, 2, ..., n$) and $E_j$ ($j = 1, 2, ..., m$) in the LP problem stated in Section 6.2. In order to minimize the computation load, it is recommended to restrict the origins and destinations to those tracts with non-zero employment and resident worker counts respectively, and compute the OD travel time $c_{ij}$ only between those tracts.

The LP problem for measuring wasteful commuting is defined by the spatial distributions of resident workers and employment and the travel time between them, which are prepared in the previous steps. These files (wrk _ job _ inBR _ 2010.txt, odd-ist.txt, and odtime.txt) are provided directly under the data folder for your convenience. The following illustrates the

---

* Again, Res represents the number of workers who both live and work in Baton Rouge, and Emp denotes the number of jobs that have the workplace and employee home addresses located in Baton Rouge. Both were extracted from CTPP part 3.

implementation of the LP approach to measure wasteful commuting in an R program.

## Step 4  Downloading and Installing R

R is a free statistical computing package and runs on several operating systems such as Windows, Mac OS, and UNIX. It can be accessed and downloaded from its mirrors (http://cran.r-project. org/mirrors.html) based on a location near you. For example, we choose to download the version "R for Windows" from the mirror from the University of California, Berkeley (http://cran.cnr. Berkeley.edu). Under the list of subdirectories (e.g., base, contrib, Rtools), choose the *base* subdirectory if it is your first time to install R. Download the current (dated July 2, 2018) version *R3.5.1 for Windows* and install it.

## Step 5  Downloading and Installing the LP Package in R

Launch R. From the main menu of R, select Packages > Install package(s) > In the CRAN mirror window, similar to step 4, choose a location close to you (in our case, we chose "USA (CA1)") for illustration. Under the list of Packages, choose *lpSolve* to install this package.

## Step 6  Running the Program for Measuring Wasteful Commuting

For a demonstration of the method, we only showcased the process to estimate wasteful commuting time. After loading the package *lpSolve*, from the main menu, select File > Open script. Navigate to the program file WasteComm _ LP.R and open it. Edit the locations of two input files (wrk _ job _ inBR _ 2010.txt and odtime.txt) and one output file (min _ comm _ time.csv) and save the script. The following R script measures the wasteful commuting at the tract level:

```
1.  library(lpSolve)
2.  #load the lpSolve library into R
```

```
3. data_od<-read.table("D:/Comm_GIS/odtime.
txt",header=T)
4. #This step reads data from odtime.txt
into a data frame
5. data_od<-data_od[order(data_od[,1],data_
od[,2]),]
6. #Sort data first based on origin tract
ID, second based on destination tract ID
7. vec_time<-data_od[,3]
8. vec_time<-t(vec_time)
9. #Then convert the third column of the
data frame (i.e., od time) to a vector
10. costs<-matrix(vec_time,nrow=92,ncol=92,b
yrow=T)
11. #Convert the above vector to a matrix
for subsequent use
12. data<-read.table("D:/Comm_GIS/wrk_job_in
BR_2010.txt",header=T)
13. #Read the constraints (numbers of
resident workers and jobs) from wrk_job_
inBR_2010.txt into a data frame
14. vec_res<-data[,2]
15. vec_res<-t(vec_res)
16. #Convert the number of resident workers
to vectors
17. vec_emp<-data[,3]
18. vec_emp<-t(vec_emp)
19. #Convert the number of jobs to vectors
20. row.signs<-rep("=",92)
21. row.rhs<-vec_res
22. col.signs<-rep("=",92)
23. col.rhs<-vec_emp
24. #Set up constraint signs and right-hand
sides
25. lp.transport(costs,"min",row.signs,row.r
hs,col.signs,col.rhs)
26. #Run to measure the minimum commuting
time
```

```
     27. result<-data.frame(data_od,as.vector
(t(lp.transport(costs,"min",row.signs,row.rh
s,col.signs,col.rhs)$solution)),row.names=NULL)
     28. colnames(result)<-c("Oid","Did","Time",
"Flow")
     29. write.csv(result,file="D:/Comm_GIS/min_c
omm_time.csv",row.names=F)
     30. #Write solution to a csv file
```

From the main menu, select Edit > Run All to run the whole program. Alternatively, one may choose the option "Run line or selection" to run the script line by line.

In the R Console window, the result is saved in a text file min _ comm _ time.csv containing the IDs of origin and destination tracts, and the travel distance and number of commuters on the route. The R Console window also shows that the objective function (total minimum commute time) is 1,207,404 minutes, that is, an average of 6.61 minutes per resident worker (with a total of 182,705 commuters who both live and work in Baton Rouge). We then replaced odtime.txt in line 3 with oddist.txt (provided in the data folder) and ran the program again to derive the optimal commuting distance. Results showed that the average minimum commuting time in Baton Rouge in 2010 was 6.61 minutes, and the average minimum commuting distance was 3.44 miles. Note that the intrazonal travel times and distances are not zero here and vary with tract area sizes. On the contrary, solving an LP model with zero intrazonal travel costs would lead to a high percentage of optimal commute trips being within the same zone. For example, 90.7 percent of the optimal commute trips were intrazonal commute in Small and Song's (1992) study. As we considered intrazonal travel costs, the optimal commuting pattern has only 24.4 percent (i.e., 44,590/182,705) within-tract commute trips when travel time is used, and 19.5 percent (i.e., 35,597/182,705) within-tract trips when distance is used.

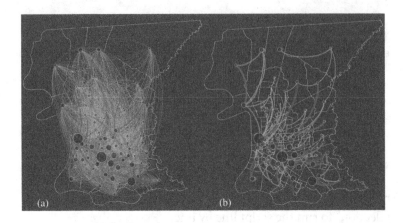

FIGURE 6.2 Tract-level commuting networks: (a) actual flow, and (b) optimal flow. (Line width is proportional to flow volume, and bubble size represents the total throughput at a tract.)

Figure 6.2 provides a visual comparison between actual and optimal commuter flows at the census tract level. Note that trips are substantially trimmed after optimization, and most are within tracts and between tracts in proximity. Given that this zonal method assumes all people living and working in a single centroid of a zone, this pattern makes sense.

Based on reported journey-to-work flow volumes and corresponding average travel time between tracts from the CTPP part 3 data, we then measure the average actual commuting time. Traditionally, it is measured based on the observed zone-to-zone travel time and flow from survey data. However, as we argued in Section 2.4 in Chapter 2, the journey-to-work survey such as the CTPP data often contains some erroneous records and respondents might not be representative of the areas they reside. For example, we detected several commute trips of several hours for traveling only a few miles: reported mean travel time from tract 36.04 to tract 11.04 in the study area was 3.7 hours (with an estimated travel distance of 5.4 miles), and 2.8 hours from tract 22 to 26.01 (with an estimated travel distance of 2.6 miles), and so on. Furthermore, as the concept of wasteful commuting emphasizes the gap between actual and optimal commute lengths, it makes more sense to also estimate actual commute time (or distance) through network analysis since the optimal commute is an estimated measure. Therefore, we propose to measure the actual commute by using estimated network travel time to rule out bias in the survey data. Both methods are described below.

### Step 7 Measuring Average Actual Commuting Time Based on Reported Time

Add `ctppTT _ Flow` to the TOC and open its attribute table. Add a new field `Flow _ TT` (as type Float) = `TT _ AllMeans` * `Flow`. Right-click on `Flow _ TT` and record the sum value (i.e., 3,770,574). Dividing this value by the total number of commuters who both live and work in Baton Rouge, 182,705, yields an average actual commuting time of 20.64 minutes. As argued previously, reported travel time from a survey could include reporting errors and respondents might not be representative of the areas they reside. After excluding the erroneous records, the average actual commuting time in 2010 was brought down to 19.26 minutes.

### Step 8 Measuring Average Actual Commuting Time Based on Estimated Network Time

Similar to steps 1–3 in this section, intrazonal travel time here is approximated as the radius of a tract's minimum bounding circle divided by a constant speed 670.56 meters/minute (i.e., 25 mph) in this case study. The total travel time between two tracts is composed of the aforementioned network time and the intrazonal time at both the origin and destination tracts. No mean travel distance is reported in the CTPP.

For comparison, we obtained the estimated travel time and distance for all commute flows and then average estimated commute time of 12.76 minutes and average estimated commute distance of 7.42 miles.

With all the above results in place, wasteful commuting was calculated according to Equation 6.1 at the census tract level.

TABLE 6.1   Summary of Actual, Optimal, and Wasteful Commuting in Baton Rouge, 2010

|  |  | Average Commuting Time (min) | Average Commuting Distance (mile) |
|---|---|---|---|
| *Census Tract Level* |  |  |  |
| Actual | Reported | 19.26 | NA |
|  | Estimated | 12.76 | 7.42 |
| Optimal |  | 6.61 | 3.44 |
| Wasteful | Reported vs. Optimal | 65.68% | NA |
|  | Estimated vs. Optimal | 48.20% | 53.64% |
| *Simulated Individual Level* |  |  |  |
| Actual | Estimated | 12.65 | 7.63 |
| Optimal |  | 4.75 | 2.71 |
| Wasteful | Estimated vs. Optimal | 62.45% | 64.48% |

*Note:* NA means not available.

Based on Table 6.1, wasteful commuting time was 65.68 percent by comparing actual time with optimal time and dropped to 48.20 percent by comparing estimated time with optimal time. In terms of distance, wasteful commuting was 53.64 percent. The extent of wasteful commuting when using estimated time (48.20 percent) is more in line with that in distance (53.64 percent) as distance is also estimated from the road network.

## 6.4   MEASURING WASTEFUL COMMUTING AT THE INDIVIDUAL LEVEL

The formulation of optimal commute at the individual level utilizes the Monte Carlo simulation of resident workers and jobs. Index the locations for simulated individual resident workers and jobs as $k$ and $l$, respectively. The total number of simulated workers is the same as that of simulated jobs, denoted by $n$. The optimization problem is

$$\text{Minimize} \sum_{k=1}^{n} \sum_{l=1}^{n} \left( c_{kl} f_{kl} \right) / n \qquad (6.11)$$

$$\text{Subject to:} \quad \sum_{l=1}^{n} f_{kl} = 1 \qquad (6.12)$$

$$\sum_{k=1}^{n} f_{kl} = 1 \qquad (6.13)$$

$f_{kl} = 1$ when a trip from $k$ to $l$ is chosen, 0 otherwise  (6.14)

where $c_{kl}$ is the estimated travel time from resident worker location $k$ to job location $l$, and $f_{kl}$ indicates whether a journey-to-work flow is chosen by the optimization algorithm (=1 when chosen and 0 otherwise). The objective function in Equation 6.11 is to minimize the average commute time of $n$ simulated commuters. Equations 6.12 and 6.13 ensure that each worker can be assigned to one unique job and vice versa. Since the variable $f$ is a binary integer, it is an integer linear programing (ILP) problem.

The result from the above ILP is the average minimal commuting, $T_r$. As explained previously, if the estimated travel time for a Monte Carlo simulated trip between two individual points $p$ and $q$ is $c_{pq}$, the average estimated commuting time $T_a$ in the simulated pattern is $T_a = \sum_{p,q} c_{pq} / n$. Since both the origins and destinations are individual points and do not involve any area configuration, both optimal commute and existing commute are estimated from the point-to-point OD trips. The approach is thus independent of the zonal effect or MAUP.

We solved the integer linear programming problem defined in Equations 6.11–6.14 for simulated individual commuters. As discussed previously, the total number of simulated commuters (or resident workers or jobs) $n$ by the Monte Carlo approach needs to be determined by balancing accuracy and computational efficiency. Based on a series of experiments with different sample sizes, we set the value of $n$ to be 3,565, which was limited by the computation of all possible OD cost matrix (i.e., 3,565*3,565 pairs) in ESRI ArcGIS 10.1 in a PC environment. Steps 1–3 in Section 4.3 in Chapter 4 illustrated the technique to derive locations of simulated individual workers (or jobs). Follow these steps to have the locations of the 3,565 simulated commuters. Step 2 in Section 6.3 discussed the process of measuring the OD cost matrix. Follow the step to obtain odtime _ indiv.txt and oddist _ indiv.txt. Next, we will elaborate on how we implement ILP to estimate wasteful commuting at the individual level. Similar to the tract level analysis, this analysis was implemented in an R program, WasteComm _ ILP.R, available in the data folder.

```
1. library(lpSolve)
2. #Load the library lpSolve
3. data_od<-read.table("D:/Comm_GIS/odtime_
indiv.txt")
4. #Read data from odtime_indiv.txt into a
data frame
5. vec_time<-data_od[,1]
6. vec_time<-t(vec_time)
7. #Convert the first column of the data
frame (i.e., od time) to a vector
8. costs<-matrix(vec_time,nrow=3565,ncol=35
65,byrow=T)
9. #Convert the above vector to a matrix for
subsequent use
10. lp.assign(costs,direction =
"min",presolve=0,compute.sens=0)
11. #Run to measure the minimum commuting
time
12. result<-data.frame(lp.assign(costs,direc
tion = "min",presolve=0,compute.sens=0)$solutio
n)
13. write.csv(result,file="D:/Comm_GIS/min_c
omm_time_indiv.csv")
14. #Write results to a csv file
```

The ILP yielded average minimum commuting time and distance at the simulated individual level as 4.75 minutes and 2.71 miles, respectively. Both are significantly lower than the optimal commute time and distance obtained at the census tract level. This difference validates the impact of zonal effect on the measure of wasteful commuting as pointed out by Hamilton (1989).

Next, we need to estimate travel time (distance) for the simulated trips that are consistent with the actual journey-to-work flow pattern. As discussed previously, the OD trips based on the Monte Carlo simulation are randomly paired individual locations of resident workers and jobs, but the total number of simulated commuters between two tracts $i$ and $j$ is capped proportionally to the actual journey-to-work volume such as $(n/N)x_{ij}$, where $n = 3,565$ and $N = 182,705$. By doing so, zonal-level trips are disaggregated into individual trips. Based on the simulation results, the average estimated commuting time and distance are 12.65 minutes and 7.63 miles, respectively. Both estimations are very close to those obtained at the census tract level. In other words, the disaggregation does not alter the estimated commuting amounts significantly, and the difference mainly lies in the minimum (required) commuting.

The visualization of networks at the simulated individual level would be too crowded to see any pattern. We aggregated both simulated flows and optimized flows into the census tract level, shown in Figure 6.3a,b, respectively. Since the simulated flows at the individual level were intentionally designed to be proportional to the actual flows, the same pattern is observed between Figures 6.2a and 6.3a, confirming that our simulation of trips worked well. In contrast to Figure 6.2b with a simpler pattern, Figure 6.3b shows the optimal commute flows at the census tract level that were aggregated back from the individual optimal pattern, which is far more complex. Individual workers are now free to swap houses for individual job locations, instead of being confined to a tract centroid for a group of workers (or a group of jobs). Therefore, the flexibility enables more choices in the optimized

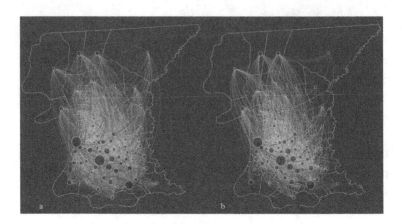

FIGURE 6.3 Aggregated individual-level commuting networks: (a) simulated flow, and (b) optimal flow. (Line width is proportional to flow volume, and bubble size represents the total throughput at a tract.)

pairing between workers and jobs, and further brings down the total (average) minimum commuting. Figure 6.3b also shows far more interzonal trips as individual workers are more likely to be paired with individual jobs in adjacent tracts instead of within the same tracts (see Figure 6.2b), and thus a more realistic optimization pattern.

Measured at the simulated individual level, we obtained 62.45 percent wasteful commuting time and 64.48 percent wasteful commuting distance. Both are higher than those estimated at the census tract level. The results are also reported in Table 6.1.

An issue regarding the sensitivity of simulation sample size merits some discussion here. The cap of $n = 3,565$ simulated commuters was set due to our computation limitation in calibrating the large $n \times n$ OD cost matrix for the ILP. Here we experimented with nine sample sizes from 820 to 3,565. Results for the average optimal commute (time and distance) and corresponding wasteful commuting percentage are shown in Figure 6.4a,b, respectively. As we increased the sample size, the average optimal commute, both in time and distance, tended to initially decrease and then stabilize after the sample size reached about 3,000.

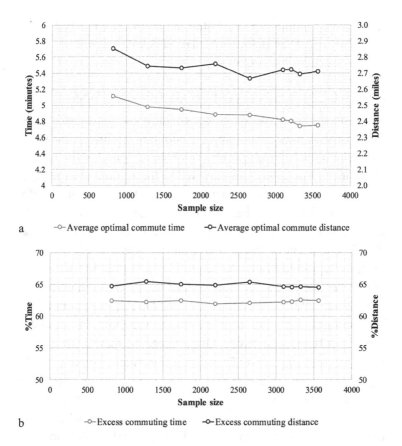

FIGURE 6.4    Sensitivity of simulation sample size: (a) optimal commute, and (b) wasteful commuting.

Even for smaller sample sizes (e.g., when *n* increased from 820 to 1,282), the changes in both measures were minor (i.e., 0.13 minutes for optimal commuting time, and 0.1 miles for optimal commuting distance). For larger sample sizes (e.g., when *n* increased from 3,256 to 3,565), the changes in the two measures were minimal (i.e., 0.004 minutes for time and 0.003 miles for distance). A similar trend can be seen on the resulting percentages of wasteful commuting: the wasteful commuting time began to converge around 62 percent and the wasteful commuting distance tended to converge around 64 percent after the sample size reached 3,000. This confirms that the sample size of 3,565 commuters was a sound choice for the study area.

## 6.5 DECOMPOSING WASTEFUL COMMUTING

Based on the results summarized in Table 6.1, Figure 6.5a illustrates various measures of average commuting time in the study area. The reported average time from survey stood at the highest value of 19.26 minutes. This may well be the amount of time experienced by commuters in general. However, the extent of wasteful commuting is derived when the actual commuting is compared to the minimal (optimal) commuting that is often based on estimation. The concept of wasteful commuting was initially proposed for the purpose of assessing the gap between what would be possible collectively for a city given its land use pattern and what individuals actually do. Therefore, it seems fairer to compare actual and optimal commute time by measuring both in estimated time from the road network. Our average estimated commuting time (driving alone) of 12.76 minutes was much shorter than the reported time of 19.26 minutes because the latter was affected by many factors such as commuters by slower modes, traffic congestion, or even mental time that survey respondents might have included to account for parking and others. As the CTPP does not report commuting distance data, this issue is not relevant when distance is used to measure commuting length.

The conventional LP approach at the zonal level yielded an average minimal (optimal) commuting time of 6.61 minutes. By simulating individual locations of resident workers and jobs and also the individual trips linking them, our research returned an average minimal commuting time of 4.75 minutes. When the zonal-level approach is used, resident workers and jobs are grouped together in zones (e.g., tract centroids), and the free swap of homes permitted by the optimization are between zones. In our approach at the simulated individual level, the optimization permits resident workers to freely swap homes with individual locations, leading to

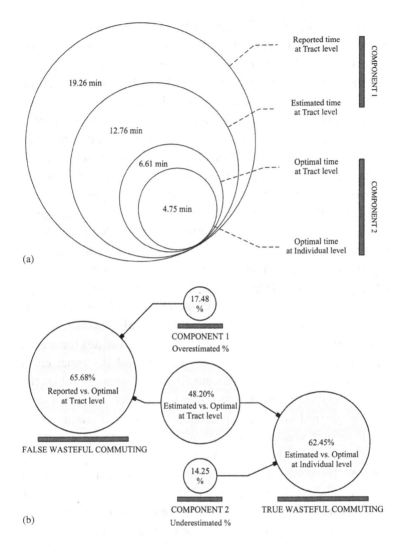

FIGURE 6.5 (a) Various measures of average commuting, and (b) decomposition of wasteful commuting.

a large number of optimal commuting trips that are much shorter and interzonal. The individual level approach is not only more accurate for its sharper resolution in estimating trip lengths but also generates the more realistic optimal commuting patterns.

Figure 6.5b further clarifies the two components of false estimation of wasteful commuting. When reported commuting time is used as a benchmark, the wasteful commuting stands as high as 65.68 percent at the zonal level. When estimated commuting time is used as a new benchmark, the wasteful commuting comes down to 48.20 percent at the zonal level. That is to say, using reported commuting time implies an overestimation of 17.48 percent wasteful commuting that is attributable to people using slower transport modes or getting caught in traffic, which a planning scenario of freely swapping house would not help. The 17.48 percent overestimation is termed "component 1" in miscalculation. By disaggregating zonal-level commuting patterns to simulated individuals, the average optimal commuting time is reduced and leads to a higher percentage of wasteful commuting (i.e., climbing back to 62.45 percent from 48.20 percent). In other words, the zonal level approach causes an underestimation of wasteful commuting by 14.25 percent, termed "component 2" in miscalculation.

In summary, the reported commuting time inflates the wasteful commuting measure by 17.48 percent, and the zonal effect underestimates it by 14.25 percent. The two components bias the estimation in opposite directions and offset each other to a large extent. The traditional approach adopted in most literature yields 65.68 percent wasteful commuting, very close to 62.45 percent obtained in our approach. However, "two wrongs do not make a right." The debate on wasteful commuting should not be about who gets the specific percentage close, rather about the scientific soundness in the means by which one reaches that percentage.

When measured in distance, the estimation error caused by component 1 is absent. The zonal level analysis generates 53.64 percent wasteful commuting. The simulated individual level analysis yields a lower average minimal distance, and thus increases the wasteful commuting to 64.48 percent, in line with the 62.45 percent wasteful commuting time.

This case study also sheds light on the choice of time or distance in measuring wasteful commuting. Some studies argue that travel time is a more appropriate estimate of travel cost and an important determining factor of travel behavior (Buliung and Kanaroglou, 2002; Gordon et al., 1991; Wachs et al., 1993). In this study, the difference in the results for time and distance is insignificant (when the estimated time is used). The only complexity is whether to use reported commuting time as a benchmark for the actual commute. If we define excessive time in using public transit or other slower modes and delayed time in congestion as "wasteful," reported time would be an adequate choice. More details about the method and analysis are reported in Hu and Wang (2015a).

## 6.6 TEMPORAL TRENDS OF WASTEFUL COMMUTING

Similar to the 2010 results listed in Table 6.1, we also measured the wasteful commuting for 1990 and 2000. Once again, the average commute distance in the study area increased steadily from 1990 to 2000 and again to 2010; and the average commute time increased from 1990 to 2000 but dropped slightly to 2010. As the concept of wasteful commuting is proposed mainly to assess the potential of commuting reduction given the land use pattern of a city, our discussion here focused on the results in terms of commute distance.

The minimum (required) commute was 3.32 miles in 1990, which dropped to 2.62 in 2000, and inched up slightly to 2.71 in 2010. It suggested that land uses in Baton Rouge might have changed in a way toward greater efficiency in terms of commuting need from 1990 to 2000 (e.g., the improved proximity between jobs and resident workers in general) and stayed largely stable until 2010. However, the resident workers did not take advantage of the change, and actually increased their trip lengths on average from 7.26 in 1990 to 7.28 in 2000 and then again to 7.42 miles in 2010. This led to the rise of wasteful commuting distance from 54.27 percent in 1990 to 63.99 percent in 2000 and stayed at 63.48 percent in 2010. Many factors may have contributed to this trend of largely increasing wasteful commuting, such as an increasing female labor participation rate (and thus more multi-worker households) and a small increase in carpool modal share from 1990 to 2000 (Table 5.1). One may refer to Hu and Wang (2016) for more discussions.

## 6.7 AN EXTENSION TO WASTEFUL COMMUTING

The assumption in measuring wasteful commuting is that people could freely exchange their homes and jobs in a city. It allows us to specifically focus on the spatial separation between homes and jobs (i.e., land use layout) by simply relocating resident workers without altering existing land use layout such as building new houses or adding new jobs. However, it becomes increasingly difficult to apply this simplistic assumption to current urban systems, where jobs (and residents) are more decentralized and social inequalities are more common. For example, it would be unfair or impossible to perform a switch between a stockbroker who lives in the suburbs and works downtown and a domestic worker in the inner city (downtown) who works in the suburbs. Besides this extreme example, there is enough specialization in society that such an assumption of equality of jobs and workers can lead to erroneous trip patterns. To make this topic more meaningful, a possible extension is to consider more factors beyond land use in the assumption. For instance, we may further constrain the relocating process to those resident workers of comparable income. Likewise, other factors such as occupation, multi-worker household, and vehicle ownership may be taken into account if such data are available so that we can obtain a more practical extent of wasteful commuting.

For illustration, we designed an example to measure wasteful commuting in Baton Rouge in 2010 by adding a constraint of comparable income in the assumption, given the data availability in CTPP (e.g., number of workers/jobs in a wage range in each tract). We first grouped resident workers and jobs into different wage strata and then repeated the above process for each wage group. For example, Equation 6.15 defines the tract-level optimal commuting for workers of a particular wage group $g$, where $x_{ij}^g$ is the number of commuters who are in wage group $g$, live in tract $i$, and work at jobs of the same wage group $g$ in tract $j$; $c_{ij}^g$ is the corresponding commuting time between them; $n^g$ is the total number of commuters of wage group $g$; and refer to Equations 6.7 through 6.9 for the constraints (note, number of workers $R_i$ is now changed to $R_i^g$ and number of jobs $E_j$ to $E_j^g$). Similarly, Equation 6.16 formulates the individual-level optimal commuting for workers of a specific wage group $g$ (refer to Equations 6.11 through 6.13 for constraints). As formulated in Equation 6.17, the overall optimal commuting for commuters of all wage groups is the weighted average of $T_r^g$ of various wage groups, where $wg$ is the number of wage groups defined.

$$\text{Tract level: } T_r^g = \min\left\{ \sum_i \sum_j \left( c_{ij}^g x_{ij}^g \right) \Big/ n^g \right\} \qquad (6.15)$$

$$\text{Individual level: } T_r^g = \min\left\{ \sum_{k=1}^{n^g} \sum_{l=1}^{n^g} \left( c_{kl}^g f_{kl}^g \right) \Big/ n^g \right\} \qquad (6.16)$$

$$\overline{T_r} = \sum_{g=1}^{wg} n^g T_r^g \Big/ \sum_{g=1}^{wg} n^g \qquad (6.17)$$

The proposed simulation process could be applied here to measure wasteful commuting of each wage group at the individual level, but its value and accuracy could be compromised without adopting additional data. Specifically, without better knowledge, a random distribution in space (e.g., tracts) is reasonable for the overall resident workers, but not necessarily applicable to subgroups of workers (e.g., wage groups) given the fact that workers of all wage groups could be present in the same census tract. For more accurate and meaningful estimates, one possible way is to define the spatial extents of each wage group (e.g., where they live and work) based on other data sources such as house price data, or to use the big data that capture the real trip information from individuals (big data and its applications in commuting studies are discussed in more detail in Section 7.3 in Chapter 7). Without these auxiliary data in place, instead, our example here focused on measures at the census tract level.

Table 6.2 reports the results for commuting time (distance is not reported). We divided workers (and jobs) into the same five groups as defined in Hu et al. (2017). Actual commuting time was measured by network travel time, and the same convex pattern that is reported in Hu et al. (2017) was observed here; it increased from the low-wage group and peaked at the medium, and then declined toward the high-wage group. Notably, the low- and high-wage group workers, on average, had a shorter commute time than the general workers, while the other three groups commuted for a longer time than the overall workers. Optimal commuting was found to be stable across wage groups at 6.61 minutes, the same as the general commuters. That is, existing land use layout in Baton Rouge in 2010 tended to be fair for workers of different wage levels, but some groups (e.g., lower-medium, medium, and upper-medium) chose to commute longer than what was suggested, and

TABLE 6.2 Wasteful Commuting Breakdowns by Wage Groups in Baton Rouge in 2010

| Wage Group | | Actual Commuting (min) | Optimal Commuting (min) | Wasteful Commuting |
|---|---|---|---|---|
| Low | <15k | 12.24 | 6.6078 | 0.46 |
| Lower-Medium | 15–35k | 12.99 | 6.6077 | 0.49 |
| Medium | 35–50k | 13.33 | 6.6103 | 0.50 |
| Upper-Medium | 50–75k | 13.04 | 6.6077 | 0.49 |
| High | >75k | 12.37 | 6.6070 | 0.47 |
| Total | | 12.76 | 6.6081 | 0.48 |

thus incurred wasteful commute. The wasteful commute was detected to vary slightly across wage groups as a result of their differentiation in actual commute. The weighted average percentage was reported to be 48 percent for workers of all wage groups, almost the same as the result reported in Table 6.1 where no comparable income assumption is appended. The diversity of resident workers/jobs in each census tract might play a role in smoothing the variation of wasteful commuting across wage groups and thus did not drive up the weighted average percentage when homogeneous wage level was imposed. In sum, even though similar percentages were obtained after adding an additional constraint, this line of research still needs more investigation—it is not the specific percentage that matters, rather the means by which one reaches that percentage.

# CHAPTER 7

# Conclusions

THIS BOOK AIMS AT developing GIS-based simulation and analytical models and applying them to commuting studies such as detecting the intra-urban commuting patterns (i.e., commute distance and time) and efficiency (i.e., wasteful commuting), and explaining the observed patterns by spatial factors such as land use. The Monte Carlo simulation method is used to measure commuting lengths at the most disaggregate level (i.e., individuals) for mitigating the area aggregation errors and scale effect, which are commonly encountered in existing studies. A major data source for our study is the Census for Transportation Planning Packages (CTPP) data (1990, 2000, and 2006–2010) from the U.S. Census Bureau. The CTPP describes commuting characteristics in three parts: Part 1 on residential places, Part 2 on workplaces, and Part 3 on journey-to-work flows. Our study area is Baton Rouge, Louisiana.

This chapter summarizes major findings from the study, highlights methodological contributions, identifies limitations, and discusses future work.

## 7.1 MAJOR FINDINGS

First, as income rises and transportation infrastructure improves over time, commuting is likely to become lengthier. This study indicates that indeed the mean commuting distance in Baton Rouge steadily increased over time, and thus workers, in general, moved farther away from their jobs in order to get better housing or maximize their earnings. Mean commuting time rose along with mean commuting distance during 1990–2000 but dipped slightly during 2000–2010. The gap between the two in 2000–2010 was attributable to a rising modal share of driving alone that was faster than other modes such as public transits.

Second, researchers in urban studies have long attempted to explain the intra-urban variability of commuting by land use patterns. It is fair to say that results from existing studies (most on large cities) have been mixed. This book reports that the mean commuting distance variation in our study area can be well explained by the distance from the central business district (CBD), jobs–housing balance ratio, and even more than 90 percent by the job proximity index. The models on mean commuting time also show improvement over existing studies. The study in Los Angles by Giuliano and Small (1993) suggested that urban land use had little impact on commuting, and their model explained very little of the actual commuting. The much-improved explaining power reported in this study can be attributable to improved measures of commute lengths, and perhaps also benefit from a moderate city size in our case. This finding lends support to the promise of planning policies that aim at trip reduction by improving the jobs–housing balance and job proximity.

Finally, the concept of wasteful commuting captures the potential for a city to reduce its overall commuting given its spatial arrangement of homes and jobs. Studies usually rely on survey data such as the CTPP to define actual commute time and measure the optimal commute at an aggregate zonal level by linear programming. This book argues that reported commute time by respondents tends to overestimate actual commute length as it includes reporting errors, travel time by slower transportation modes (e.g., public transit, bicycling, walking, and others), and delayed time due to congestion. The zonal level analysis of optimal commute also suffers from the scale effect. The former tends to over estimate the actual commuting length and leads to overestimating wasteful commuting; and the latter (especially the use of large areal unit) gives rise to overestimated optimal commuting length and leads to underestimating wasteful commuting. The two may cancel each other and thus conceal the problem. This book proposes to measure wasteful commuting at the most disaggregate level by simulating trip ends, that is, individual resident workers (O) and individual jobs (D) within zones (e.g., census tracts), and thus has effectively mitigated the scale effect. It also computes estimated commute distance and time for actual journey-to-work trips as a new benchmark for existing commuting. The resulting estimates of wasteful commuting are largely consistent between measures in time and distance.

A temporal analysis on wasteful commuting changing pattern in this book indicates that Baton Rouge in its entirety experienced an increase of wasteful commuting from 1990 to 2000 in both commute distance and time and stayed at about the same levels toward 2010. This indicates that the land use configuration changed in such a way that jobs were collectively moved closer to residences and thus became better balanced from 1990 to 2000, but the resident workers did not take advantage of that and incurred more wasteful commuting. The trend of rising wasteful commuting was largely halted between 2000 and 2010. The economic downturn beginning in 2008 might be one reason underlying the new trend (Horner and Schleith, 2012). The low temporal resolution data (i.e., 5-year pooled 2006–2010 CTPP) prevent us from validating this speculation.

## 7.2 METHODOLOGICAL CONTRIBUTIONS AND LIMITATIONS

This book proposes a Monte Carlo simulation method to measure commuting lengths more accurately by first simulating individual resident workers and jobs that are consistent with their spatial distributions across the areal unit (e.g., census tract), and then simulating individual trips that are proportional to the existing area-based journey-to-work trip flows. It is a significant improvement over the zonal-level centroid-to-centroid approach and can mitigate the aforementioned aggregation error and scale effect. In addition, wasteful commuting, specifically the optimal commuting, is now reformulated as an integer linear programming approach targeted on individual trip makers rather than the traditional linear programming approach on the collective behavior and patterns from the overall commuters.

There are also some limitations to our study. First, future research could expand the study area by incorporating neighboring counties (parishes) to see whether there are more pronounced changes in the metropolitan area. If high temporal resolution data in recent years become available, one may be able to examine the impact of major external factors (e.g., significant fluctuation in gas prices) on commuting. Future work can also include other transportation modes such as public transit, biking, and walking to fully capture the urban mobility for all socio-demographic groups as this study focused only on personal vehicles for daily commutes. Furthermore, this book follows existing research in measuring wasteful commuting by assuming that people can freely swap houses for minimal commutes. Future research could relax this assumption by limiting swaps to matches in income, occupation, and household characteristics, which are more realistic (refer to Section 6.7 in Chapter 6, for example). Last, it would be beneficial to additionally examine the link between nonspatial factors, such as the wage rate discussed in Hu et al. (2017), and commuting patterns.

## 7.3 BIG DATA, BETTER STORIES

A key approach used in this study is the Monte Carlo simulation that generates individual resident workers and jobs, and individual trips linking them, all of which are simulated in a way consistent with those reported in the census data in an area unit (here, census tracts). The purpose of our simulation is to improve the estimation of trip lengths (distance and time) between points instead of between area centroids. In other words, the simulation approach is to overcome the limitation of census data, a common source for commuting and many other socioeconomic studies. However, simulated trips are not real (though close to real trips, at least in a way that is consistent with what has been recorded in the census).

Where do we find data on individual commute trips? The answer lies in the availability of "big data," especially human motion data from location aware devices (LADs) including mobile phones and Global Positioning System (GPS) receivers. In contrast to "designed data" such as the census data collected via surveys, big data are "organic" and collected automatically by tracking transactions of all sorts (Groves, 2011). Therefore, big data are normally composed of a large volume of individual trips. One major challenge of using big data for commuting analysis is the lack of identification of trip purposes in the data. One may approximately detect commuting trips by analyzing the spatio-temporal pattern of such data in conjunction with some traditional data sources (e.g., land use data and field surveys) (Liu et al., 2012). When it is feasible to prepare a subset of commuting trips from big data, it has several major advantages. The data are (1) individual trips with accurate records of trip origins, destinations, and trip lengths (avoiding the reporting errors of census data discussed in Section 6.3), (2) representation of a large population (if including groups on various transportation modes), (3) updated instantly, and (4) cost-efficient. Studies on the interdependence between commuting and land use and the extent of wasteful commuting, as reported in this book, will be replicated with more confidence. In short, the data are sharper, faster, and cheaper, and thus enable us to tell better stories (Wang, 2018). However, the paucity of socio-demographic attributes associated with trip makers, a common limitation of big data, would prevent one from explaining commute patterns by nonspatial factors as discussed in Section 2.3.

The authors recently purchased big data of individual vehicular trips that took place in the month of September 2017 in the Tampa Bay area (Hillsborough County and Pinellas County) in Florida. The data were collected through smartphones, GPS, and other sensors installed in cars or on the road. For each trip, the locations (recorded in latitude and longitude), dates, and times of the origin, destination, and intermediate waypoints were recorded. Certainly, such rich information has the potential to be applied in a wide range of urban studies including commuting. For example, we can summarize the data to detect the locations of homes and workplaces of each individual trip maker and then analyze the commuting patterns (actual and optimal patterns). As the recorded locations in the data are the actual places where trips occurred but not simulated ones, corresponding analyses showcased in this book (and beyond) would be more accurate and the findings could be more beneficial to the decision makers. In addition to improved spatial resolution, such big data can also benefit studies interested in detecting temporal variability of travel patterns such as daily or hourly (in)stability. Both spatial and temporal dimensions are important to geographic research and other social sciences as well. We will report the results from analyzing these data in the near future.

# References

Aguilera, A. 2005. Growth in commuting distances in French polycentric metropolitan areas: Paris, Lyon and Marseille. *Urban Studies* 42(9): 1537–47.

American Association of State Highway and Transportation Officials (AASHTO). 2013. *Commuting in America 2013: The National Report on Commuting Patterns and Trends*. Washington, DC: American Association of State Highway and Transportation Officials.

American Association of State Highway and Transportation Officials (AASHTO). 2014. CTPP data product based on 2006–2010 5-year American Community Survey (ACS) data. https://ctpp.transportation.org/ctpp-data-set-information/5-year-data/ (accessed July 29, 2018).

Anselin, L. 1995. Local Indicators of Spatial Association U.S.—LISA. *Geographical Analysis* 27: 93–115.

Antipova, A., Wang, F., and Wilmot, C. 2011. Urban land uses, sociodemographic attributes and commuting: A multilevel modeling approach. *Applied Geography* 31(3): 1010–18.

Besag, J., and Diggle, P. J. 1977. Simple Monte Carlo tests for spatial pattern. *Applied Statistics* 26: 327–33.

Boussauw, K., Van Acker, V., and Witlox, F. 2012. Excess travel in nonprofessional trips: Why look for it miles away? *Tijdschrift voor Economische en Sociale Geografie* 103(1): 20–38.

Buliung, R. N., and Kanaroglou, P. S. 2002. Commute minimization in the Greater Toronto area: Applying a modified excess commute. *Journal of Transport Geography* 10(3): 177–86.

Bureau of Transportation Statistics (BTS). 2014. Census Transportation Planning Package (CTPP). 1990 CTPP http://www.transtats.bts.gov/tables.asp?db_id=620&DB_Name and 2000 CTPP http://www.transtats.bts.gov/tables.asp?DB_ID=630 (accessed July 29, 2018).

Cervero, R. 1989. Jobs-housing balancing and regional mobility. *Journal of the American Planning Association* 55(2): 136–50.

Cervero, R. 1996. Jobs-housing balance revisited: Trends and impacts in the San Francisco Bay area. *Journal of the American Planning Association* 62(4): 492–511.

Cervero, R., Chapple, K., Landis, J., Wachs, M., Duncan, M., Scholl, P. L., and Blumenberg, E. 2006. Making do: How working families in seven U.S. metropolitan areas trade off housing costs and commuting times. Research report, Institute for Transportation Studies, University of California, Berkeley. https://cloudfront.escholarship.org/dist/prd/content/qt9wf8x6p5/qt9wf8x6p5.pdf (accessed July 29, 2018).

Cervero, R., and Wu, K. L. 1998. Sub-centring and commuting: Evidence from the San Francisco Bay area, 1980–90. *Urban Studies* 35(7): 1059–76.

Charron, M. 2007. From excess commuting to commuting possibilities: More extension to the concept of excess commuting. *Environment and Planning A* 39(5): 1238–54.

Chen, X., Zhan, F. B., and Wu, G. 2010. A spatial and temporal analysis of commute pattern changes in Central Texas. *Annals of GIS* 16(4): 255–67.

Clark, C. 1951. Urban population densities. *Journal of Royal Statistical Society* 114: 490–4.

Clark, W. A., Huang, Y., and Withers, S. 2003. Does commuting distance matter?: Commuting tolerance and residential change. *Regional Science and Urban Economics* 33(2): 199–221.

Clifford, P., Richardson, S., and Hémon, D. 1989. Assessing the significance of the correlation between two spatial processes. *Biometrics* 45: 123–34.

Cooke, T., and Marchant, S. 2006. The changing intrametropolitan location of high-poverty neighbourhoods in the U.S., 1990–2000. *Urban Studies* 43 (11): 1971–89.

Crane, R., and Chatman, D. G. 2003. Traffic and sprawl: Evidence from U.S. commuting, 1985 to 1997. *Planning and Markets* 6(1): 14–22.

Cropper, M. L., and Gordon, P. L. 1991. Wasteful commuting: A re-examination. *Journal of Urban Economics* 29(1): 2–13.

Downs, A. 1992. *Stuck in Traffic: Coping with Peak-Hour Traffic Congestion.* Washington, DC: Brookings Institution; Cambridge, MA: Lincoln Institute of Land Policy.

Dubin, R. 1991. Commuting patterns and firm decentralization. *Land Economics* 67(1): 15–29.

Fan, Y., Khattak, A., and Rodríguez, D. 2011. Household excess travel and neighbourhood characteristics: Associations and trade-offs. *Urban Studies* 48(6): 1235–53.

Fields, A., and Jiles, M. E. 2009. Workers who drove alone to work: 2007 and 2008 American Community Surveys. American Community Survey Reports, U.S. Department of Commerce, Economics and Statistics Administration, U.S. Census Bureau. https://www.census.gov/prod/2009pubs/acsbr08-5.pdf (accessed July 29, 2018).

Fisher, R. A. 1935. *The Design of Experiments*. Edinburgh: Oliver and Boyd.

Frankena, M. W. 1978. A bias in estimating urban population density functions. *Journal of Urban Economics* 5(1): 35–45.

Frost, M., Linneker, B., and Spence, N. 1998. Excess or wasteful commuting in a selection of British cities. *Transportation Research Part A: Policy and Practice* 32(7): 529–38.

Gao S., Wang Y., Gao Y., and Liu Y. 2013. Understanding urban traffic-flow characteristics: A rethinking of betweenness centrality. *Environment and Planning B: Planning and Design* 40(1): 135–53.

Gera, S. 1979. Age and the journey-to-work: Some further empirical evidence. *Socio-Economic Planning Sciences* 13(5): 285–87.

Gera, S., and Kuhn, P. 1980. Job location and the journey-to-work: An empirical analysis. *Socio-Economic Planning Sciences* 14(2): 57–65.

Giuliano, G., and Small, K. A. 1993. Is the journey to work explained by urban structure? *Urban Studies* 30(9): 1485–500.

Gordon, P., Kumar, A., and Richardson, H. W. 1989a. Congestion, changing metropolitan structure, and city size in the United States. *International Regional Science Review* 12(1): 45–56.

Gordon, P., Kumar, A., and Richardson, H. W. 1989b. Gender differences in metropolitan travel behaviour. *Regional Studies* 23(6): 499–510.

Gordon, P., Kumar, A., and Richardson, H. W. 1989c. The influence of metropolitan spatial structure on commuting time. *Journal of Urban Economics* 26(2): 138–51.

Gordon, P., Lee, B., and Richardson, H. W. 2004. *Travel Trends in U.S. Cities: Explaining the 2000 Census Commuting Results*. Los Angeles: Lusk Center for Real Estate, University of Southern California. https://lusk.usc.edu/sites/default/files/working_papers/wp_2004-1007.pdf (accessed July 29, 2018).

Gordon, P., Richardson, H. W., and Jun, M. J. 1991. The commuting paradox evidence from the top twenty. *Journal of the American Planning Association* 57(4): 416–20.

Groves, R. 2011. "Designed Data" and "Organic Data." https://www.cen sus.gov/newsroom/blogs/director/2011/05/designed-data-and-or ganic-data.html (accessed July 29, 2018).

Hamilton, B. W. 1982. Wasteful commuting. *Journal of Political Economy* 90(5): 1035–53.

Hamilton, B. W. 1989. Wasteful commuting again. *Journal of Political Economy* 97(6): 1497–504.

Hewko, J., Smoyer-Tomic, K. E., and Hodgson, M. J. 2002. Measuring neighbourhood spatial accessibility to urban amenities: Does aggregation error matter? *Environment and Planning A* 34(7): 1185–206.

Hitchcock, F. L. 1941. The distribution of a product from several sources to numerous localities. *Journal of Mathematics and Physics* 20(2): 224–30.

Horner, M. W. 2002. Extensions to the concept of excess commuting. *Environment and Planning A* 34(3): 543–66.

Horner, M. W. 2004. Spatial dimensions of rban commuting: A review of major issues and their implications for future geographic research. *Professional Geographer* 56(2): 160–73.

Horner, M. W. 2007. A multi-scale analysis of urban form and commuting change in a small metropolitan area (1990–2000). *Annals of Regional Science* 41(2): 315–32.

Horner, M. W., and Murray, A. T. 2002. Excess commuting and the modifiable areal unit problem. *Urban Studies* 39(1): 131–9.

Horner, M. W., and Schleith, D. 2012. Analyzing temporal changes in land-use–transportation relationships: A LEHD-based approach. *Applied Geography* 35(1): 491–8.

Hu, Y., and Wang, F. 2015a. Decomposing excess commuting: A Monte Carlo simulation approach. *Journal of Transport Geography* 44: 43–52.

Hu, Y., and Wang, F. 2015b. Monte Carlo method and its application in urban traffic simulation. In *Quantitative Methods and Socio-Economic Applications in GIS*, ed. F. Wang, 259–277. Boca Raton, FL: CRC Press.

Hu, Y., and Wang, F. 2016. Temporal trends of intraurban commuting in Baton Rouge, 1990–2010. *Annals of the American Association of Geographers* 106(2): 470–9.

Hu, Y., Wang, F., and Wilmot, C. G. 2017. Commuting variability by wage groups in Baton Rouge, 1990–2010. *Papers in Applied Geography* 3(1): 14–29.

Hu, Y., Wang, F., Guin, C., and Zhu, H. 2018. A spatio-temporal kernel density estimation framework for predictive crime hotspot mapping and evaluation. *Applied Geography* 99: 89–97.

Huff, D. L. 1963. A probabilistic analysis of shopping center trade areas. *Land Economics* 39(1): 81–90.

Ikram, S. Z., Hu, Y., and Wang, F. 2015. Disparities in spatial accessibility of pharmacies in Baton Rouge, Louisiana. *Geographical Review* 105(4): 492–510.

Jin, S., Yang, L., Danielson, P., Homer, C., Fry, J., and Xian, G. 2013. A comprehensive change detection method for updating the National Land Cover Database to circa 2011. *Remote Sensing of Environment* 132: 159–75.

Kain, J. F. 1968. Housing segregation, negro employment, and metropolitan decentralization. *Quarterly Journal of Economics* 82(2): 175–97.

Kim, S. 1995. Excess commuting for two-worker households in the Los Angeles metropolitan area. *Journal of Urban Economics* 38(2): 166–82.

Kim, C. 2008. Commuting time stability: A test of a co-location hypothesis. *Transportation Research Part A: Policy and Practice* 42(3): 524–44.

Kneebone, E., and E. Garr. 2010. *The Suburbanization of Poverty: Trends in Metropolitan America, 2000 to 2008*. Washington, DC: Brookings Institution: Metropolitan Policy Program. https://www.brookings.edu/wp-content/uploads/2016/06/0120_poverty_profiles.pdf (accessed July 29, 2018).

Kwan, M. P., and Kotsev, A. 2015. Gender differences in commute time and accessibility in Sofia, Bulgaria: A study using 3D geovisualisation. *Geographical Journal* 181(1): 83–96.

Layman, C. C., and Horner, M. W. 2010. Comparing methods for measuring excess commuting and jobs-housing balance. *Transportation Research Record: Journal of the Transportation Research Board* 2174(1): 110–7.

Levinson, D. M. 1998. Accessibility and the journey to work. *Journal of Transport Geography* 6(1): 11–21.

Levinson, D. M., and Kumar, A. 1994. The rational locator: Why travel times have remained stable. *Journal of the American Planning Association* 60(3): 319–32.

Levinson, D., and Wu, Y. 2005. The rational locator reexamined: Are travel times still stable? *Transportation* 32(2): 187–202.

Li, M., Kwan, M. P., Wang, F., and Wang, J. 2018. Using points-of-interest data to estimate commuting patterns in central Shanghai, China. *Journal of Transport Geography* 72: 201–210.

Liu, Y., Wang, F., Xiao, Y., and Gao, S. 2012. Urban land uses and traffic 'source-sink areas': Evidence from GPS-enabled taxi data in Shanghai. *Landscape and Urban Planning* 106(1): 73–87.

Lowe, K., and Marmol, M. E. 2013. Worker experiences of accessibility in post-Katrina New Orleans. UNOTI Publications. Paper 16. https://scholarworks.uno.edu/unoti_pubs/16 (accessed July 29, 2018).

Luo, W., and Wang, F. 2003. Measures of spatial accessibility to health care in a GIS environment: Synthesis and a case study in the Chicago region. *Environment and Planning B: Planning and Design* 30(6): 865–84.

Luo, L., McLafferty, S., and Wang, F. 2010. Analyzing spatial aggregation error in statistical models of late-stage cancer risk: A Monte Carlo simulation approach. *International Journal of Health Geographics* 9: 51.

Ma, K. R., and Banister, D. 2006. Excess commuting: A critical review. *Transport Reviews* 26(6): 749–67.

Massey, D. S., Rothwell, J., and Domina, T. 2009. The changing bases of segregation in the United States. *Annals of the American Academy of Political and Social Science* 626(1): 74–90.

McKenzie, B. 2014. Modes less traveled—Bicycling and walking to work in the United States: 2008–2012 (No. ACS-25). U.S. Department of Commerce, Economics and Statistics Administration, U.S. Census Bureau. https://www.census.gov/prod/2014pubs/acs-25.pdf (accessed July 29, 2018).

McKenzie, B., and Rapino, M. 2011. Commuting in the United States: 2009. U.S. Department of Commerce, Economics and Statistics Administration, U.S. Census Bureau. https://www.census.gov/prod/2011pubs/acs-15.pdf (accessed July 29, 2018).

Merriman, D., Ohkawara, T., and Suzuki, T. 1995. Excess commuting in the Tokyo metropolitan area: Measurement and policy simulations. *Urban Studies* 32(1): 69–85.

Mills, E. S. 1972. *Studies in the Structure of the Urban Economy*. Baltimore: Johns Hopkins University.

Murphy, E., and Killen, J. E. 2011. Commuting economy: An alternative approach for assessing regional commuting efficiency. *Urban Studies* 48(6): 1255–72.

Muth, R. 1969. *Cities and Housing*. Chicago: University of Chicago.

National Household Travel Survey. 2006. Commuting for life. https:// nhts.ornl.gov/briefs/Commuting%20for%20Life.pdf (accessed July 29, 2018).

Niedzielski, M. A., Horner, M. W., and Xiao, N. 2013. Analyzing scale independence in jobs-housing and commute efficiency metrics. *Transportation Research Part A: Policy and Practice* 58: 129–43.

Niedzielski, M. A., O'Kelly, M. E., and Boschmann, E. E. 2015. Synthesizing spatial interaction data for social science research: Validation and an investigation of spatial mismatch in Wichita, Kansas. *Computers, Environment and Urban Systems* 54: 204–18.

O'Kelly, M. E., and Lee, W. 2005. Disaggregate journey-to-work data: Implications for excess commuting and jobs-housing balance. *Environment and Planning A* 37(12): 2233–52.

Openshaw, S., and Taylor, P. J. 1979. A million or so correlation coefficients: Three experiments on the modifiable areal unit problem. In *Statistical Applications in Spatial Sciences*, ed. N. Wrigley, 127–44. London: Pion.

Peng, Z. R. 1997. The jobs-housing balance and urban commuting. *Urban Studies* 34(8): 1215–35.

Pisarski, A. 2002. *Testimony for "Mobility, Congestion and Intermodalism."* Washington, DC: U.S. Senate Committee on Environment and Public Works.

Poulter, S. R. 1998. Monte Carlo simulation in environmental risk assessment: Science, policy and legal issues. *Risk: Health, Safety and Environment* 9: 7–26.

Roberto, E. 2008. Commuting to opportunity: The working poor and commuting in the United States. https://www.brookings.edu/rese arch/commuting-to-opportunity-the-working-poor-and-comm uting-in-the-united-states/ (accessed July 29, 2018).

Robinson, W. S. 1950. Ecological correlations and the behavior of individuals. *American Sociological Review* 15: 351–7.

Rodríguez, D. A. 2004. Spatial choices and excess commuting: A case study of bank tellers in Bogota, Colombia. *Journal of Transport Geography* 12(1): 49–61.

Rosenbloom, S., and Burns, E. 1993. *Gender Differences in Commuter Travel in Tucson: Implications for Travel Demand Management Programs.* Berkeley, CA: University of California Transportation Center. https://escholarship.org/uc/item/036776w2 (accessed July 29, 2018).

Ross, M., and Svajlenka, N. P. 2012. Connecting to opportunity: Access to jobs via transit in the Washington, DC region. https://www.brookings.edu/research/connecting-to-opportunity-access-to-jobs-via-transit-in-the-washington-d-c-region/ (accessed July 29, 2018).

Santos, A., McGuckin, N., Nakamoto, H. Y., Gray, D., and Liss, S. 2011. Summary of Travel Trends: 2009 National Household Travel Survey (No. FHWA-PL-ll-022). https://nhts.ornl.gov/2009/pub/stt.pdf (accessed July 29, 2018).

Scott, D. M., Kanaroglou, P. S., and Anderson, W. P. 1997. Impacts of commuting efficiency on congestion and emissions: Case of the Hamilton CMA, Canada. *Transportation Research Part D: Transport and Environment* 2(4): 245–57.

Shen, Q. 2000. Spatial and social dimensions of commuting. *Journal of the American Planning Association* 66(1): 68–82.

Shen, Y., Kwan, M. P., and Chai, Y. 2013. Investigating commuting flexibility with GPS data and 3D geovisualization: A case study of Beijing, China. *Journal of Transport Geography* 32: 1–11.

Shi, X. 2009. A geocomputational process for characterizing the spatial pattern of lung cancer incidence in New Hampshire. *Annals of the Association of American Geographers* 99(3): 521–33.

Small, K. A., and Song, S. 1992. "Wasteful" commuting: A resolution. *Journal of Political Economy* 100(4): 888–98.

Suh, S. H. 1990. Wasteful commuting: An alternative approach. *Journal of Urban Economics* 28(3): 277–86.

Sultana, S. 2002. Job/housing imbalance and commuting time in the Atlanta metropolitan area: Exploration of causes of longer commuting time. *Urban Geography* 23(8): 728–49.

Sultana, S., and Weber, J. 2007. Journey-to-work patterns in the age of sprawl: Evidence from two midsize southern metropolitan areas. *Professional Geographer* 59(2): 193–208.

Sultana, S., and Weber, J. 2014. The nature of urban growth and the commuting transition: Endless sprawl or a growth wave? *Urban Studies* 51(3): 544–76.

Taaffe, E. J., Gauthier, H. L., and O'Kelly, M. 1996. *Geography of Transportation* (2nd ed.). Upper Saddle River, NJ: Prentice Hall.

Taylor, G. E. 1989. Addendum to Saalfield (1987). *International Journal of Geographical Information Systems.* 3(2): 192–3.

Texas Transportation Institute. 2011. 2011 Urban Mobility Report. https://static.tti.tamu.edu/tti.tamu.edu/documents/ums/archive/mobility-report-2011-wappx.pdf (accessed July 29, 2018).

Thurston, L., and Yezer, A. M. 1991. Testing the monocentric urban model: Evidence based on wasteful commuting. *Real Estate Economics* 19(1): 41–51.

U.S. Census Bureau. 2012. U.S. Census Bureau, Statistical Abstract of the United States: 2012, Table 7. https://www.census.gov/library/publications/2011/compendia/statab/131ed.html (accessed July 29, 2018).

U.S. Department of Transportation/Federal Highway Administration. 2012. Highway Statistics 2012, Table VM-202. http://www.fhwa.dot.gov/policyinformation/statistics/2012/vm202.cfm (accessed July 29, 2018).

U.S. Environmental Protection Agency. 2018. Sources of Greenhouse Gas Emissions. https://www.epa.gov/ghgemissions/sources-greenhouse-gas-emissions (accessed July 29, 2018).

Wachs, M., Taylor, B. D., Levine, N., and Ong, P. 1993. The changing commute: A case-study of the jobs-housing relationship over time. *Urban Studies* 30(10): 1711–29.

Wang, F. 2000. Modeling commuting patterns in Chicago in a GIS environment: A job accessibility perspective. *Professional Geographer* 52(1): 120–33.

Wang, F. 2001. Explaining intraurban variations of commuting by job proximity and workers' characteristics. *Environment and Planning B: Planning and Design* 28(2): 169–82.

Wang, F. 2003. Job proximity and accessibility for workers of various wage groups. *Urban Geography* 24(3): 253–71.

Wang, F. 2015. *Quantitative Methods and Socio-Economic Applications in GIS*. Boca Raton, FL: CRC Press.

Wang, F. 2018. "Big Data, Better Stories." Speech delivered on June 26 at the 35th Hongmen Forum on Spatiotemporal Big Data and Future Cities, Peking University, Beijing, China.

Wang, F., Antipova, A., and Porta, S. 2011. Street centrality and land use intensity in Baton Rouge, Louisiana. *Journal of Transport Geography* 19(2): 285–93.

Wang, F., Hu, Y., Wang, S., and Li, X. 2017. Local indicator of colocation quotient with a statistical significance test: Examining spatial association of crime and facilities. *Professional Geographer* 69(1): 22–31.

Wang, F., Liu, C., and Xu. Y. 2018. Mitigating the zonal effect in modeling urban population density functions by Monte Carlo simulation. *Environment and Planning B: Urban Analytics and City Science*. DOI: 10.1177/2399808317749832.

Watanatada, T., and Ben-Akiva, M. 1979. Forecasting urban travel demand for quick policy analysis with disaggregate choice models: A Monte Carlo simulation approach. *Transportation Research Part A* 13: 241–8.

Wegener, M. 1985. The Dortmund housing market model: A Monte Carlo simulation of a regional housing market. *Lecture Notes in Economics and Mathematical Systems* 239: 144–91.

White, M. J. 1988. Urban commuting journeys are not "wasteful." *Journal of Political Economy* 96(5): 1097–110.

Yang, J. 2008. Policy implications of excess commuting: Examining the impacts of changes in U.S. metropolitan spatial structure. *Urban Studies* 45(2): 391–405.

Yang, J., and Ferreira, J. 2008. Choices versus choice sets: A commuting spectrum method for representing job-housing possibilities. *Environment and Planning B: Planning and Design* 35(2): 364–78.

Zax, J. S., and Kain, J. F. 1991. Commutes, quits, and moves. *Journal of Urban Economics* 29(2): 153–65.

Printed in the United States
by Baker & Taylor Publisher Services